THE
DOCTOR

Also by
DR KARL KRUSZELNICKI

Curious & Curiouser
Brain Food
50 Shades of Grey Matter
Game of Knowns
House of Karls
Dr Karl's Short Back & Science

Dinosaurs Aren't Dead
Dr Karl's Big Book of Science Stuff (and Nonsense)
Dr Karl's Even Bigger Book of Science Stuff (and Nonsense)
Dr Karl's Biggest Book of Science Stuff (and Nonsense)
Dr Karl's Big Book of Amazing Animals

THE DOCTOR

DR KARL KRUSZELNICKI

Pan Macmillan Australia

First published 2016 in Macmillan by Pan Macmillan Australia Pty Ltd

1 Market Street, Sydney, New South Wales, Australia, 2000

Copyright © Karl S. Kruszelnicki Pty Ltd 2016

The moral right of the author has been asserted.

All rights reserved. No part of this book may be reproduced or transmitted by any person or entity (including Google, Amazon or similar organisations), in any form or by any means, electronic or mechanical, including photocopying, recording, scanning or by any information storage and retrieval system, without prior permission in writing from the publisher.

Cataloguing-in-Publication entry is available
from the National Library of Australia
http://catalogue.nla.gov.au

Cover, internal design and typesetting by Xou Creative, www.xou.com.au

Cover photograph of Dr Karl by Mel Koutchavlis
Other cover images: iStock

Internal illustrations by Roy Chen, Xou Creative

Heat wave guidelines on page 39 reproduced by permission, NSW Ministry of Health © 2016.

Printed by McPherson's Printing Group

The author and the publisher have made every effort to contact copyright holders for material used in this book. Any person or organisation that may have been overlooked should contact the publisher.

The publishers and their respective employees or agents will not accept responsibility for injuries or damage occasioned to any person as a result of participation in the activities described in this book. It is recommended that individually tailored advice is sought from your healthcare professional.

Papers used by Pan Macmillan Australia Pty Ltd are natural, recyclable products made from wood grown in sustainable forests. The manufacturing processes conform to the environmental regulations of the country of origin.

I dedicate this book to the Theoretical Scientists. They need full and reliable funding (and definitely without the "Pub Test", in My Universe).

They dive into the Unknown, to give us WiFi, GPS Satellite Navigation, and Laser-Readers at our supermarkets. They love their work so much that they work over the Christmas Holidays. Which is when, on 25 December 1990, they invented the World Wide Web (I wonder if it will ever take off?) and then gave it to the world – for free.

They discovered that protons live inside atoms. That research translated into the KnowHow to devise a beam of protons that deliver their energy exactly into the cancer – not on the skin, not deeper in the flesh, but exactly into the cancer.

I can't wait to see what practical uses that will come (one day) from **Dark Energy**, **Dark Matter** and **Gravitational Waves**. So let the Minds of Theoretical Scientists run free over the Vast Expanses of Space and Time.

TABLE OF CONTENTS

01. ALCOHOL AND HEARING 1

02. PASTEURISED MILK GOES OFF? 9

03. DOGS TILT HEAD 14

04. VAMPIRE BLOOD DRINKING 17

05. PERPETUAL PRESENT 24

06. PERPETUAL PAST 26

07. HEAT WAVE 29

08. ROADS OF ICE 40

09. REFUEL CAR WITH ENGINE RUNNING? 43

10. GRAVITATIONAL WAVES 49

11. THE HEIGHT OF GOOD HEALTH 67

12. PYTHON THE CRUSHER 75

13. FLY EYES AND SOLAR PANELS 83

14. CEMENT SHOES 90

15. HOW MANY CELLS IN YOUR BODY? 93

16. WATER BURNS PLANTS? 103

17. DIRTY DATA 111

18. IMMORTAL JELLYFISH 119

19. HOT TEA COOLS YOU DOWN 122

20. TIME TRAVEL 125

21. BITCOIN: LEGEND OF A LEDGER 129

22. SPLEEN AND RED BLOOD CELLS 150

23. MOVIE AUDIENCES EMIT CHEMICALS 153

24. MOZZIES LOVE (SOME) HUMANS 157

25. DOUBLE-YOLK EGGS 161

26. GOD, CAFFEINE & CHOCOLATE 164

27. CREDIT CARD THEFT 167

28. SLEEP BADLY IN AN UNFAMILIAR BED 177

29. SMOOTHIE SCAM 180

30. SUNSCREEN ATTACKS CORAL REEF 183

31. COFFEE'S A DIURETIC? 188

32. DRAGONFLY TELESCOPE 191

33. PHUNDAMENTAL PHYSICS PROBLEMS 199

34. PLANET – A GIANT BABY 205

35. POISONED PANTS 211

36. I, VOMIT 218

37. VOMITING MACHINE 221

38. PLACES OF PI 229

39. ELECTRIC MOTORS IN BACTERIA 245

40. COFFEE IS NOW GOOD 251

41. WEIGHT LOSS VIA EXERCISE 260

REFERENCES 263

ACKNOWLEDGMENTS 285

01

ALCOHOL AND HEARING

IT'S HARD TO PINPOINT EXACTLY. BUT AT EVERY PARTY, THERE COMES A STAGE WHEN THE MOOD PICKS UP.

People have drunk enough alcohol to loosen their inhibitions, start relaxing and dancing. The observable link to the effect of the alcohol is that the noise level goes up with the mood.

Why does the party get louder when people drink more alcohol? What is alcohol doing to your brain, and what are the flow-on effects?

ALCOHOL 101

Alcohol, known as ethyl alcohol, or C_2H_5OH, has wondrous properties. It can remove oil stains from the garage floor, and store body parts or axolotls beautifully for centuries.

And in humans, drinking small quantities of this versatile liquid can improve mood and self-confidence, as well as get the conversation flowing. But in bigger doses, the effects are less wondrous. Alcohol then interferes with your fine muscle control and your higher mental functioning, which messes with your decision-making.

ALCOHOL EFFECTS

The legal driving level of alcohol in the blood varies around the world. It's usually measured as the Blood Alcohol Concentration, or BAC. It's just a ratio – so 0.05 per cent BAC means that one twentieth of one per cent of your total blood volume is alcohol. The Legal Driving BAC is zero per cent in some countries such as Afghanistan, Indonesia, Hungary and Nepal, 0.02 per cent in China, Sweden and Norway, 0.05 per cent in Australia, New Zealand and many European countries, and 0.08 per cent in the USA, England and Wales.

With a BAC at the lower end (0.03 per cent to 0.12 per cent) you get the desired effects of mild euphoria and general improvement in mood. These go hand-in-hand with lower anxiety, and increased self-confidence and sociability. You still get adverse side effects at low BAC, including a flushed red face, and a minor loss of both fine muscle coordination and mental judgement.

By the time you get to BAC levels between 0.09 per cent and 0.25 per cent, you're looking at blurred vision, problems with balance, and drowsiness. As you head up to around 0.30 per cent BAC, you're dizzy, staggering, vomiting, slurring your words and dangerously confused. Things get really nasty around 0.40 per cent BAC, with life-threatening respiratory depression and unconsciousness. At higher BAC levels, coma and even death from alcohol poisoning are on the cards.

FEEDBACK LOOP

But back to party mode, why that sudden increase in noise?

We're still not entirely sure, but it seems to involve a feedback loop. Let me explain. Once you have a few drinks, your sense of hearing is impaired. So when you speak, you mistakenly think that you are talking more softly than usual. To compensate, you start talking louder.

We should also factor in Social Competition and Loss of Inhibition. As people try to get their Words of Wisdom across, there is a bit of "shouting down the competition". This is called the Lombard Effect (after Étienne Lombard, a French otolaryngologist, who discovered it in 1909).

YOU GO SLIGHTLY DEAF

Interestingly, alcohol's effects on hearing are different for men and women. Women go deafer than men.

In most studies, men and women (in a Double Blind situation) drank juice, either with or without alcohol. Once the people drinking alcohol hit around 0.03 per cent BAC, their hearing was tested at six different audio frequencies running from low to high (250, 500, 1000, 2000, 4000 and finally 8000 hertz). On average, the men would lose 2–9 decibels of hearing, while women would lose more, 5–12 decibels. A lot of the hearing is lost around the 500–1000 hertz range – which are the frequencies where a lot of speech happens, and where vowels are discriminated.

---

1) Internal fridge environment is low humidity → drying of moist food items. 2) Cover stops transfer of smell b/w food items.

Unfortunately, in most of the studies, the sample sizes are rather small so statistics are less reliable. From this limited data, the trend seemed to be that women, fat people, unhealthy people, older people, and those with a history of heavier drinking lost more of their hearing in each drinking event. Luckily, within a week of drinking, the hearing tended to return to pre-alcohol–drinking levels. Overall, most of the temporary hearing loss was down at the lower frequencies.

But the hearing loss was different for long-term drinkers. They tended to have permanent hearing loss, and more often, at the higher frequencies.

We're not sure why.

RISING VERSUS FALLING BAC

It turns out that your performance can be different at the same BAC, depending on whether the BAC is rising (you are getting more drunk) or falling (heading for sobriety). This has been measured for tasks such as muscle coordination, cognitive skills and how accurate you are at rating your level of drunkness.

In general – and this goes for humans and some non-humans – you perform worse on the way up. On the way down, some time has passed, and you've had the chance to adapt. Pharmacologists call this behaviour "Acute Tolerance".

DUNNO WHERE IT HAPPENS

Sound information is carried from your eardrum to the central processing centres inside your brain.

How does alcohol make you slightly deaf? *Where* in the hearing chain does alcohol make you slightly deaf? We have some tantalising hints from the relatively few studies done, but the simple answer is that we don't know.

This hearing loss might be from the alcohol having a direct toxic effect on the cochlea, or a subtle influence on neurotransmission, or a direct anaesthetic effect or osmotic effect on nerves or hair cells in the cochlea, or changes in impedance of the moving bones of the inner ear... or something else. We don't really know.

Where does it happen? The alcohol might be acting on your eardrum, or the muscles that can pull on the eardrum to quieten down the outside world, or the cochlea, or the Cochlear Nerve that carries the information into your brain, or it could be acting on the area in your brain that processes this information.

Regardless of the exact pathway by which it happens, the result once you've had a few drinks is that you "hear" yourself as if you are talking too quietly. And to compensate, you start talking loudly.

UNFORTUNATE DEAFNESS

There are many causes of hearing loss in our modern industrial world. Social drinking in the evening for most people usually happens in a noisy bar. This adds to the noise-induced hearing loss the partygoers experience in everyday life.

We worry about getting blind drunk. But maybe another concern for the inebriated is getting deaf drunk...

All glass is under stress. Might be caused by repeated heating/cooling stress cycles from items on top of glass, sunlight, etc. → more local stress.

SOUND TO SENSE

I started my university education in Physics and Maths. It felt good knowing how the Universe started with the Big Bang. But I had no idea about the Physiology of the human body – I just thought my insides were some kind of chunky red salsa that could leak out through the skin if I got injured.

When I started studying Biomedical Engineering, my very first assignment was the human ear. I was totally amazed by how pressure waves in the air get turned into "words" in your "consciousness".

First, a friend says "hello" to you.

Second, pressure waves commonly called "sound" travel through the air from their mouth to your ear.

Third, these pressure waves are somewhat focused by your external ear to travel down your ear canal. As they travel down the tapered diameter of your ear canal, their "impedance" is changed to better match the "impedance" of the eardrum. (You can think of impedance as being a "resistance to being moved".)

@DoctorKarl Are there earthquakes on the Moon?

Fourth, when you listen to the quietest sound that you can hear, your eardrum moves back and forth a tiny distance roughly equal to the diameter of a hydrogen atom.

Fifth, the moving eardrum sends its energy through two tiny bones (the hammer and anvil) into a third tiny bone (the stirrup). These bones move very freely, because they have the smallest synovial joints (the Rolls-Royce of joints) in the body. They also have tiny, fast-acting muscles that can pull on the bones to dampen very loud sounds, such as thunder.

Sixth, the stirrup pushes on a flexible membrane called the Oval Window, belonging to a structure called the cochlea, which has internal hair cells. On the other side of the Oval Window is a special liquid bathing these hair cells. The pressure waves move through this liquid. They force the "hairs" on hair cells to bend. When the hairs bend, they give off electricity.

Seventh, this electricity is picked up by the Cochlear Nerve and sent to the brainstem. It travels to various stations in the brainstem and brain until it ends up at the Auditory Cortex. The electricity is now decoded in the Wernicke's Area, and you finally "hear" the word "hello".

And all of this happens in just a tiny fraction of a second.

Yes (rarely). Even more rare & mysterious are lights on the Moon (no, not UFOs).

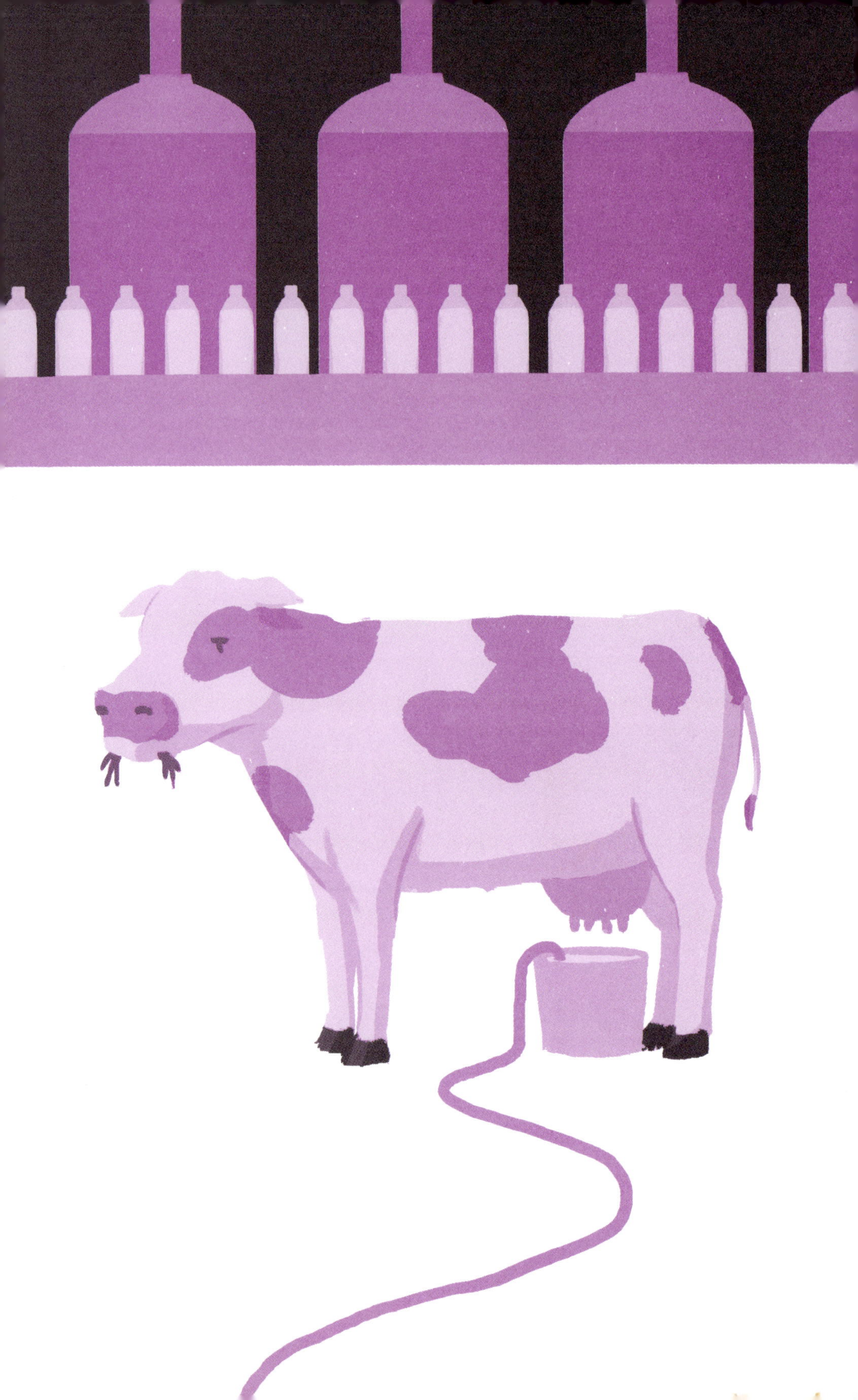

02

PASTEURISED MILK GOES OFF?

IF PASTEURISATION IS SUPPOSED TO KILL ALL THE GERMS IN RAW MILK, WHY DOES PASTEURISED MILK GO OFF AFTER A WEEK?

You may have seen "pasteurised" on the side of your milk carton. Pasteurisation is a heat process that kills most of the bacteria in liquid foods – "most" not "all". Pasteurisation aims to kill 99.999 per cent of the microbes (yeasts, moulds, bacteria and the like) – this is called the "Five Nines" Standard. The remaining microbes are so few in number that you've got about a week to a fortnight after pasteurisation before they multiply enough to cause a problem.

Pasteurisation was originally invented by a French chemist, Louis Pasteur, to treat wine, not milk. On a summer holiday in the town of Arbois in 1864, he found that the local winemakers had a problem with their wines turning acid. Pasteur discovered he could kill microbes in wine by heating it to about 55°C for a few minutes. Bingo, the wines no longer turned acid. But this took decades to be applied to milk. In the USA, routine pasteurisation of raw milk began in the 1920s, and was widespread by 1950. Pasteurisation was introduced to Australia in the late 1950s.

Milk, by an unfortunate coincidence, is one of the best culture mediums for growing bacteria. Raw milk has not been pasteurised, and is crawling with bacteria. In the US between 1998 and 2011, thanks to raw milk and raw cheese products, there were 284 hospitalisations and two deaths. Raw milk killed a three-year-old Melbourne child in December 2014.

Regular pasteurisation can be done by heating the milk either to around 60°C for about 20 minutes, or about 72°C for about 15 seconds. The nutritional losses are insignificant – calcium and phosphorus drop by 5 per cent, vitamins B_1 and B_{12} by 10 per cent, and vitamin C by 20 per cent.

Alternatively, the Ultra Heat Treating (UHT) process involves heating the milk to about 140°C for four seconds. This should kill all bacteria. If the UHT milk is then packed under sterile conditions into a pre-sterilised airtight container, it should last for nine months. The high-temperature sets off the Maillard Reaction, which alters the taste

@DoctorKarl Why does hyperthermia cause heart attacks?

and flavour. The structure of the milk proteins is also changed, so UHT milk is not really suitable for making cheese. Most other nutrients are at similar levels to regularly pasteurised milk except for folate, which drops by about 90 per cent.

Raw onions only make you cry, but raw milk might make you die.

Enzymes (essential for all biochemical reactions, including Cardiac Muscle contraction) work ONLY within narrow temp range.

RAW MILK Q&A

DOESN'T RAW MILK HAVE SPECIAL ENZYMES THAT KILL BAD BACTERIA?

NO. THE OPPOSITE IS TRUE. BACTERIA LOVE TO GROW IN MILK. ACCORDING TO THE US CENTERS FOR DISEASE CONTROL AND PREVENTION, RAW MILK IS ONE OF THE WORLD'S RISKIEST FOOD PRODUCTS. BEFORE WIDESPREAD PASTEURISATION IN ENGLAND AND WALES, BETWEEN 1912 AND 1937 ABOUT 65,000 PEOPLE DIED FROM TUBERCULOSIS CONTRACTED FROM MILK.

DOESN'T RAW MILK HAVE SPECIAL ENZYMES SO THAT YOUR BODY CAN DIGEST IT BETTER AND GET MORE NUTRITION FROM IT?

NO.

IF THE MILK-PRODUCING ANIMAL (COW, GOAT, ETC.) IS CLEAN AND HEALTHY, AND RAISED IN SANITARY CONDITIONS, DOESN'T THAT MAKE THE MILK FREE OF BACTERIA?

NO. AN ANIMAL CAN BE PERFECTLY HEALTHY, AND STILL CARRY BACTERIA THAT DON'T AFFECT THAT ANIMAL. BUT THAT CAN AFFECT US. ONCE THE BACTERIA GET INTO THE MILK, THEIR NUMBERS INCREASE VERY RAPIDLY. (AS AN ASIDE, WE'VE RECENTLY DISCOVERED THAT BATS CAN CARRY MANY VIRUSES THAT CAN KILL HUMANS, BUT DON'T AFFECT THE BATS AT ALL.)

IF THE RAW MILK IS GENUINELY AND PROPERLY CERTIFIED "ORGANIC", IS THE MILK SAFE TO DRINK?

NO. "ORGANIC" HAS MANY DEFINITIONS ACROSS THE WORLD. AT THE VERY LEAST, IT IMPLIES THAT THE FOOD IS FREE OF INSECTICIDES AND THE LIKE. THIS IS FINE. BUT "ORGANIC" MILK CAN STILL BE LOADED WITH HARMFUL BACTERIA. HOWEVER, IT IS POSSIBLE TO BUY PASTEURISED ORGANIC MILK.

@DoctorKarl Anti-ageing Gin – for real?

IS RAW MILK FROM ANIMALS THAT ARE FED GRASS, NOT GRAIN, PERFECTLY SAFE?
NO. BACTERIA ARE EVERYWHERE. HARMFUL BACTERIA HAVE BEEN FOUND IN GRASS-FED ANIMALS. THEY ARE OF THE SAME TYPES, AND AT THE SAME LEVELS, AS IN GRAIN-FED ANIMALS. ONE PLUS IS THAT THE ECOLOGICAL WATER LOAD ON THE ENVIRONMENT IS CERTAINLY LESS WITH GRASS-FED ANIMALS.

CAN RAW MILK CURE OR PREVENT DISEASES SUCH AS CANCER, HEART DISEASE, ALLERGIES AND ASTHMA?
NO. HOWEVER, IMMUNE SYSTEM CONDITIONS SUCH AS ALLERGIES AND ASTHMA CAN SOMETIMES BE VERY SENSITIVE TO CERTAIN CHEMICAL TRIGGERS IN THE ENVIRONMENT. IT CAN BE WORTHWHILE TRYING DIFFERENT PASTEURISED MILKS.

CHEESES AND YOGHURTS MADE FROM RAW MILK HAVE BEEN PROCESSED MORE. ARE THEY SAFE TO EAT?
NO. IN FACT, THE TWO DEATHS IN THE USA BETWEEN 1998 AND 2011 CAUSED BY RAW MILK AND ITS PRODUCTS WERE CAUSED BY A FRESH MEXICAN-STYLE CHEESE MADE FROM RAW MILK.

An expensive alcoholic drink that prevents ageing? Did that claim appear in 1) an advertisement 2) peer-reviewed journal?

03 DOGS TILT HEAD

People will often chat to their dogs. When they start talking, the dog sometimes tilts its head to one side in a most lovable manner.

Stanley Coren, author of many dog books, thought about this.

One explanation was that the dog wanted an ear to be aimed at their owner. Another was that it was a canine social cue – they wanted us to know that they were paying attention.

But Stanley Coren thought of a more obvious answer – the dog's long snout or muzzle was blocking their view of their owner's mouth. Perhaps they were tilting their head to get the muzzle out of the way.

@DoctorKarl Why do your eyes water when you choke on your food?

Try this experiment. Make a fist. Hold it up to your nose. Look at a fellow human. You can't clearly see their mouth (the centre of many facial expressions, and where words come from). Then tilt your head to one side – now you can see their mouth.

It's a nice hypothesis, but how would you prove it?

Stanley realised that some dogs have rather flat faces. So he carried out an online survey. (Mind you, this was a self-reported survey, so this makes the results less trustworthy.) A large number of people responded – 582 dog owners.

Stanley compared dogs with flatter faces to "regular" dogs with longer muzzles. Seventy-one per cent of the dogs with larger muzzles tilted their heads when spoken to, as compared to 52 per cent of dogs with flatter faces. OK, so dogs with muzzles that blocked their view of their owners' faces were more likely to tilt their heads. But what about the remaining 29 per cent of dogs with longer muzzles? Don't they care about reading emotions on their owners' faces?

Maybe seeing the owner's face is one factor out of many. Or maybe the head-tilting dogs just like to look cute …

1) Dunno 2) Choking → intra-thoracic pressure → intra-skull pressure → pressure on some lacrimal glands → tears? 3) Anybody?

04

VAMPIRE BLOOD DRINKING

WHEN MOVIE-MAKERS WANT TO PLAY IT SAFE AND GUARANTEE THEIR INVESTMENT DOLLARS, THEY'LL PICK AN OLD FAVOURITE FOR THE LEAD ROLE. THAT'S WHY SANTA CLAUS, JESUS CHRIST, SHERLOCK HOLMES AND DRACULA THE VAMPIRE APPEAR SO FREQUENTLY.

Most vampire movies use the same lore. I love movies, but I also love accuracy. So sadly I have to tell you that as far as vampires are concerned, the movies tell you two big bite-sized lies.

First, it would take more than just a few seconds for the victim to become unconscious. And second, if biting the victim turns them into a vampire – well, by now we should all be vampires.

VAMPIRE 101

For a vampire to survive, we're told they have to feed on the life essence of a living creature – usually their blood. In this unnatural process, the vampire gets their nourishing meal of blood, and the victim falls unconscious – and is set on the pathway of becoming another vampire.

Legends of supernatural creatures that survive on the blood (or flesh) of the living have been around for thousands of years. The ancient Persians, Babylonians, Hebrews, Assyrians, Greeks and Romans each had their own folklore about blood-guzzling demons and spirits. Similar legends exist across Africa, both North and South America, and Asia. But it was the 1897 novel *Dracula*, by Bram Stoker, that seems to have established the modern European vampire character and subsequent Vampire Culture.

VAMPIRE – THE WORD

We're not sure where the word "vampire" came from. One of its first usages in the English language was in 1745, in a travelogue called *Travels Of Three English Gentlemen*. About a century later, the vampire concept was familiar enough for Karl Marx to describe Capital as being "dead labour which, vampire-like, lives only by sucking living labour . . ."

These days, vampires are an essential part of our TV and movie landscape.

@DoctorKarl **Is a drop of wee just water, urine or both?**

VAMPIRE SCIENCE

In 2016, a group of physics students from the University of Leicester in the UK tried to work out how long it would take an actual vampire to make a real person unconscious from a bite on the neck.

First, they assumed you would fall unconscious once you lost about 15 per cent of your blood. Now a typical (and worthy) blood donation is around 10 per cent of your blood volume. But the physics students figured that a bit more blood loss (an extra 5 per cent) combined with the shock of somebody else's canine teeth stabbing into the carotid artery in your neck would be enough to make you faint. The students also assumed that the vampire's fangs made two tiny puncture holes each 0.5 millimetres across, and that the victim's blood would come out under a pressure of around 100 millimetres Hg.

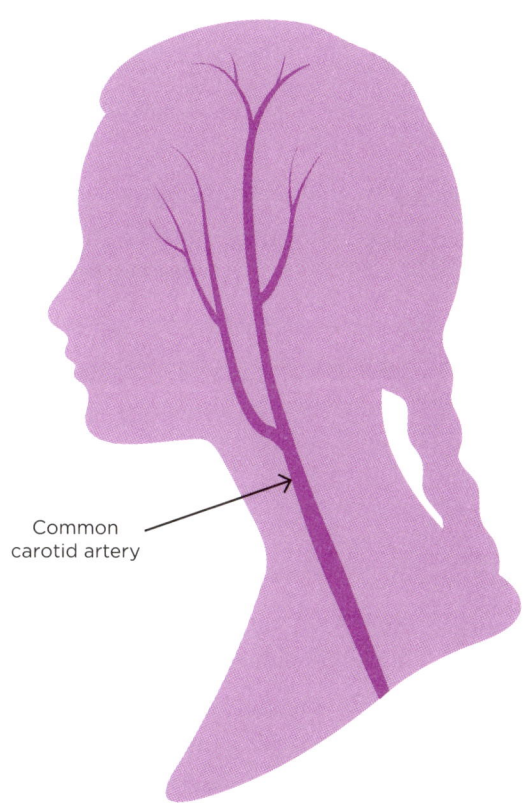

Common carotid artery

UNLESS you have retrograde flow up urethra in shower/bath THEN it's all urine from your bladder.

VAMPIRES HAVE PORPHYRIA?

Back in the mid-1980s, two scientists (Professor David Dolphin, a biochemist from the University of British Columbia, and Professor Alan Johnson from Sussex University) proposed that all the obvious symptoms of being a vampire could be explained by a fairly rare, and hereditary, blood disease called Porphyria. These included sensitivity to sunlight, drinking

Using these numbers, they came up with a time of 6.4 minutes to drain 0.75 litres of blood from the carotid artery. (By the way, the reason that it takes longer at the Blood Bank to drain about half a litre of blood is that it's seeping from a vein at low pressure, rather than squirting from an artery at high pressure.)

PROBLEM 1: NO TIME

But in the Land of Movies, the victim collapses within a few seconds. That supposedly gives the vampire enough time to drink their fill and make a quick getaway – all before the hero arrives with their crucifix and garlic. Here's where the physics students' research kicks in. It turns out that a few seconds is not enough time for the vampire to get a proper feed.

But enough of the Cardiovascular Physiology involved in Vampire Dietetics. What about the Mathematics of Vampire Enrolment and Recruitment?

Let's start with three assumptions.

The first one is very conservative – let's assume that vampires need a feed only once every month.

The second one is kind of obvious, but it has to be stated. If you've been bitten and turned into a vampire, you are no longer alive – you are "undead". That means you are no longer a potential source of nourishment for other vampires.

@DoctorKarl Big curry week – now my farts smell so bad. Why?

blood (of course), and the power of garlic to deter vampires.

It turns out that this was a tongue-in-cheek work, but I got tricked. (I've always been a sucker for a "sciency" explanation.) I wrote the story "Vampires Suck" for my second book, *Even Greater Moments in Science*. I hereby retract the story.

The third one is also obvious. Every time you add one vampire to the population, you simultaneously subtract a human.

You can see where I'm heading.

What happens when there are no more humans left to bite?

"REAL" VAMPIRES

Even as recently as this century, innocent people have been attacked on the suspicion of being blood-sucking vampires.

In December 2002, in Malawi, one man suspected of being a vampire was stoned to death. In the Thyolo region three travelling Roman Catholic priests were also beaten and detained. They were released only after a woman recognised one of them as a priest.

PROBLEM 2: NO FOOD

Begin with just one vampire in a world human population of about 7.5 billion. After one month, they plunge their fangs into the neck of a victim, who shortly joins them in the Vampire Club.

Now we have two vampires. In the third and fourth months, we double and double again to four and then eight vampires.

By 10 months, there are 512 vampires. Twenty months gets us to half a million vampires, while at 30 months we have half a billion

---

Curry + associated foods → changes in gut bacteria → different "gases" being produced → magnificently odourous "wind".

vampires roaming the planet looking for their next meal from a human. Thirty-three months after the first vampire had their first bite, there are now four billion vampires – and somewhat fewer than four billion rather nervous humans remaining.

And bingo, 34 months – less than three years – after the first vampire had their first warm, liquid meal, every human on the planet has been turned into a vampire.

Suddenly, there is no food remaining and the vampires are left with the combination of immortality and an unquenchable thirst.

But one way to protect the human stock would be to stake (i.e. kill by stabbing a wooden stake into their heart) all newly created vampires. Yes, the Vampire Cycle of Life would continue, but wouldn't they then be all sad and lonely with no vampire friends?

Vampires aren't frightening – they're only in the movies. It's the terrible Mathematical Mistakes that are absolutely ghastly...

SESAME STREET AND SEEDS

Sesame Street, the popular children's TV show, got it right with its friendly vampire, Count von Count. He has a compulsion to count everything he comes across – accidentally teaching numbers and counting to children in the process. This habit is called "arithmomania".

It also ties in perfectly with ancient techniques used in Europe and China to ward off vampires. These methods relied on vampires' unstoppable arithmomania. The local folk would place poppy seeds, millet, sand or bags of rice around the graves of presumed vampires or their own houses. Once the vampire saw the huge number of seeds or grains, they would have to stop and count every single one – and be occupied for eternity, or at least until the Sun came up (whichever came sooner).

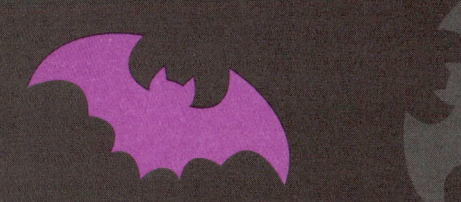

VAMPIRE ERRATA

Unfortunately, when I went to University, there were no courses on vampires, so I may have made a few mistakes. Even recently, there are also deplorably few peer-reviewed papers by Forensic Pathologists on this topic. Luckily, a few experts have been kind enough to set me straight.

First, the so-called Vampire Movies are really factually accurate documentaries.

Second, vampire teeth can transform from skinny little needles into fat fangs. So the physics students' assumption that the puncture hole was only 0.5 millimetres could be wrong. The documentaries show them as closer to 2–3 millimetres.

Third, mostly the vampires bite the jugular vein (which has low pressure), but sometimes they go for the carotid artery (at high pressure). Perhaps the physics students chose the carotid artery, because it made their maths simpler.

Fourth, becoming a vampire – the technical term is "turning" – is not always automatic. Sometimes the victim just dies and is gone forever. Sometimes they become a vampire only if they both were bitten *and* did Bad Stuff during their life. And sometimes the biting vampire will be very picky about whom they choose to anoint with the privilege of being turned into a vampire. Very occasionally, the newly created vampire has to drink the blood of the vampire who bit them to complete the cycle and turn into a vampire.

Fifth, sometimes the biting vampire will neither kill the victim, nor turn them into a fellow vampire. In this scenario, they make repeated follow-up visits to harvest more blood.

Finally, in some cases, the vampire bite carries with it some fast-acting General Anaesthetic. So that means there's no need to cause unconsciousness by draining 15 per cent of the blood volume.

Not quite. "Fabric" or "continuum" of the Universe is made of 4 Dimensions of Space–Time.

05 PERPETUAL PRESENT

Without memories, it would be hard to remember who we are.

But there are at least three people living now – that we know of – who cannot recall any memories of their past. Each is a healthy, high-functioning adult. One (known as AA, a retirement specialist) is married, one (CC, with a PhD) is in a relationship, and one (BB) is single. Their memory gaps are not due to trauma nor degenerative diseases – they were born this way (apologies to Lady Gaga). The condition is called Severely Deficient Autobiographical Memory (SDAM).

It's not that they have no memory at all. They can learn, and remember, the history of World War II, the fall of the Berlin Wall, the death of Princess Diana. But they cannot lay down memories that integrate what they feel, with time and location details, in a movie-like

@DoctorKarl Can tiredness make symptoms worse?

fashion. You and I can remember how delightful the ocean felt on our last holiday when we dived into it. We can retrieve and relive the feelings of the warmth and the saltiness of the moment. But these three with SDAM cannot.

AA can remember that she recently walked onto the stage to sing an old English folk song. But she has no idea how she felt on stage. She quickly forgets an argument and cannot hold a grudge – because she can't remember what caused it. She can't remember any books she's read or movies she's seen – each re-reading or re-watching is effectively her first experience of it, every time.

Certain parts of the brain are critical for us to lay down memories so that we can "time travel" into our past to re-experience it – and perhaps to understand ourselves better. These regions include the Left Medial Prefrontal Cortex (responsible for mental projections of one's self back through time) and the Right Precuneus (visual memory). The three SDAM sufferers have reduced brain activity in these regions.

Some of us who ruminate on the past (perhaps more than we need) might think it would be liberating to live only in the here and now – effortlessly, in a state of eternal mindfulness.

But the rest of us probably love all our little golden memories, even if they are inaccurate and hazy – and wouldn't want to give them up for the world.

Very much so. As parents know, sleep can be better than sex . . .

06
PERPETUAL PAST

There are some people who can remember most days of their lives as though they happened just a month ago. They have Highly Superior Autobiographical Memory (HSAM).

If you ask them what happened on June-the-somethingth, 15 years ago, they'll tell you it was a Wednesday, and that the day started off sunny, but there was a storm in the afternoon. On their way home, the train got delayed because a tree had fallen across the railway tracks. And sure enough, the public record agrees with their private recollection.

The first person diagnosed with HSAM emailed a psychologist in the year 2000. Today, some 50 people with this quirk of memory have been documented.

As compared to you and me, they have the same recall for events and

@DoctorKarl **Do you love what you do and is it good for you?**

TODAY YESTERDAY TODAY

personal experiences up to about a month ago. But then, as the months and years and decades roll by, their recall stays fairly constant, whereas ours relentlessly fades away.

Often, their HSAM emerged around the age of eleven.

In most cases, these individuals have a strong tendency to obsessive behaviours. They might have to wash their car keys with alcohol if they fall to the ground. They might keep a very detailed diary, or just before they go to sleep, they might pick a day at random and try to remember what happened on that day in successive years. They are more likely to fantasise about events that relate to themselves, and they easily get absorbed in new experiences.

They are slightly better than the rest of us when tested for recall of visual objects, and associating names with faces.

But in other aspects of their lives, they are completely average. They might have five keys on their key ring, but not be able to say what each key is for.

Surprisingly, they are also slightly more prone to false memory. For example, in a test run by Patihis et al., they might recall seeing footage of a news event for which no footage at all exists.

On their brain scans, two areas that are related to memory are larger than a normal person's – the Uncinate Fascicle and the Parahippocampal Gyrus.

I wonder what happens after an argument. Sure, they can forgive, but what about forget?

I love writing when I get "in the zone" → world around me fades away.
Health? Do exercise, write standing & sitting.

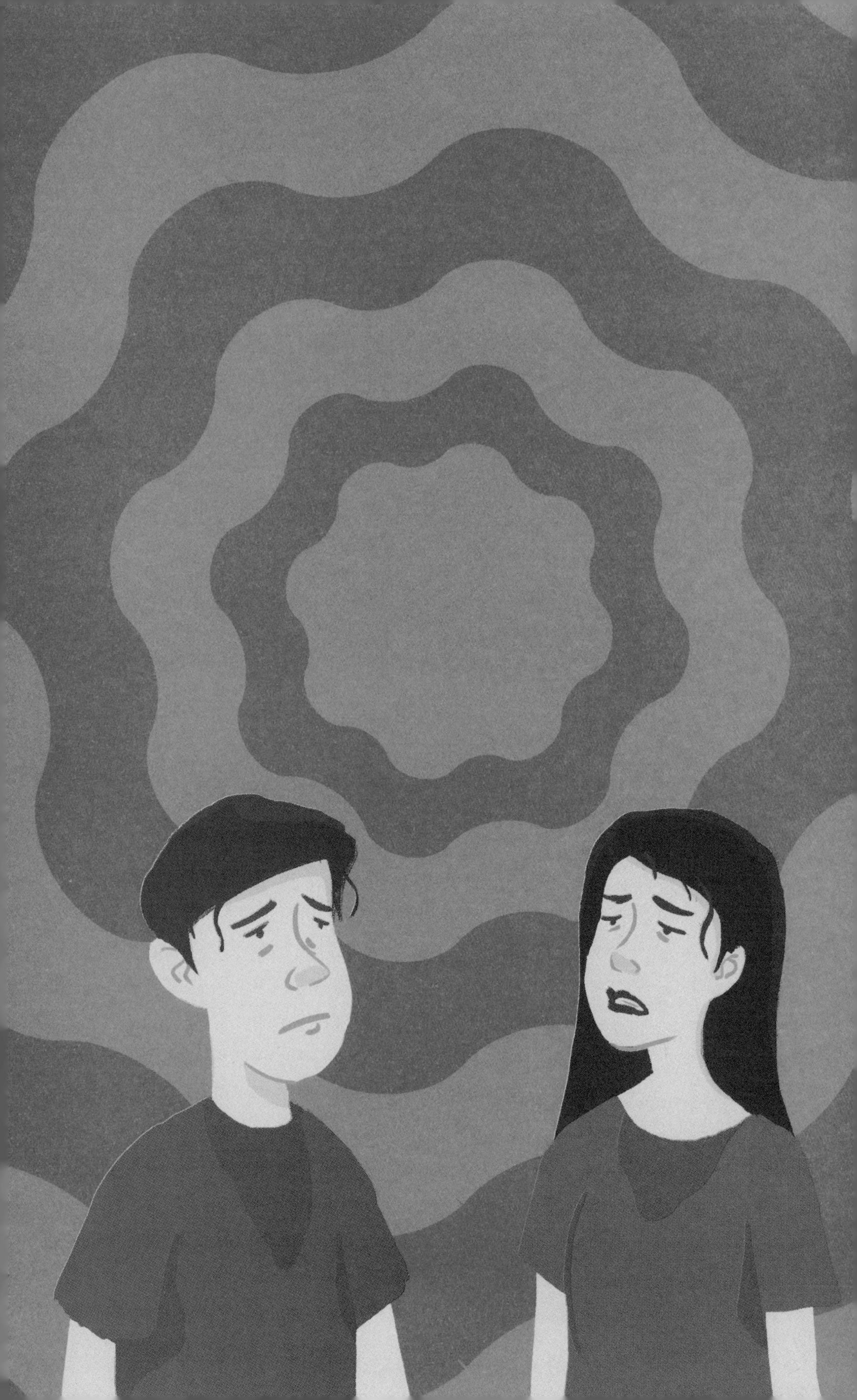

07

HEAT WAVE

HEAT WAVES KILL MORE PEOPLE IN AUSTRALIA THAN BUSHFIRES DO. BY 2040, IN EACH CALENDAR YEAR, HEAT WAVES WILL SWEEP ACROSS 20 PER CENT OF THE LAND AREA OF OUR PLANET. (THAT'S A MASSIVE INCREASE FROM JUST 1 PER CENT BACK IN THE 1960S.)

Most Australians would remember the terrible 2009 Black Saturday disaster in Victoria. The flames, heat and smoke from those horrendous bushfires killed 173 people. But what most Australians don't realise is that the crippling heat waves around Black Saturday killed 374 people. That's more than twice as many people.

YES, HEAT WAVES CAN KILL

Overall, heat waves have killed more Australians than all other natural hazards combined – 55 per cent of all such deaths. More than 4500 Australians have died from heat waves since the year 1900.

In the European heat wave of 2003, some 50,000–70,000 people died between June and August. The Russian heat wave of 2010 killed around 55,000 people.

Thanks to Global Warming, future heat waves will be more extreme in temperature, happen more frequently, last longer, and cover more of the Earth's surface.

HEAT WAVE: MORE THAN 5°C FOR FIVE DAYS

First, what exactly is a heat wave?

Confusingly, the definition varies depending on the country. Sometimes it can vary from one state to another within a country – such as in the USA.

One widely accepted definition comes from the World Meteorological Organization (WMO).

The WMO starts by setting the baseline as the 30 years between 1961 and 1990. Then, for the location you're interested in, pick one day of the year, e.g. 27 January. Each of those "27 January" days across the 30 years will have a minimum and a maximum temperature (usually after midnight, and after midday). Add up the 30 maximum temperatures, divide by 30, and then you have the Average Maximum Temperature for that particular day and location. Do the same for the next four days to give you the average maximum temperature for five days in a row.

According to the WMO, a heat wave happens when you have five days in a row that each have a daily maximum temperature five or more Celsius degrees higher than the Average Maximum Temperature.

CAUSE OF HEAT WAVES

Second, what causes a heat wave?

Basically, a heat wave occurs when a high-pressure system in the atmosphere, instead of moving across the landscape, stays stuck in one location – for days or even weeks.

But in the mega heat waves that killed tens of thousands of people in Europe and Russia, there was another factor. These heat waves were made worse by a vicious positive feedback loop between ultra-dry soil and unexpectedly powerful high-pressure systems in the lowest level of the atmosphere. This combination trapped the heat. The trapped heat couldn't dissipate overnight – so the next morning started off as hot as the previous afternoon. The cycle intensified with each passing day. Ultimately, it created a thick blanket of hot air, four kilometres thick.

RECOGNISE HEAT WAVE DEATHS

Third, how can you tell if a specific death is caused by a heat wave?

Mostly, you can't – directly. There are many factors involved. How well you tolerate heat depends on what temperatures you are used to, your age and general health, your home's architecture and its location, and so on. So the heat wave that kills one person might not kill their neighbour. Furthermore, when you do an autopsy, there is no specific pathology in the corpse that implicates a heat wave as the cause of death.

But you can tell, indirectly, when heat waves cause deaths. You know that something very bad is happening when the dead bodies start to pile up in the morgue.

In the heat waves of Europe in 2003, Victoria in 2009, Russia in 2010, and Victoria again in 2014, the morgues had the metaphorical "No Vacancy" signs up. There was simply no more room to store the dead bodies that were coming in. The overflow had to be stored in mortuaries, universities and funeral parlours. In Paris in 2003, the corpses of most of the 15,000 heat-related victims had to be stored temporarily in a refrigerated warehouse outside the city.

Earth accelerates around Sun → emits Gravitational Waves → gets closer to Sun each year one-third of trillionth of metre.

Then, after the heat wave has passed, you call in the statisticians to work out how many people it killed. They compare the number of deaths during the heat wave with the number of deaths over the same time period in previous years.

In Australia, the most lethal day for a heat wave is the day after Australia Day – 27 January.

CAUSE OF HEAT DEATH

What exactly is it that kills somebody in a heat wave?

Amazingly, we still don't fully understand what goes on. In Paris alone in 2003, some 15,000 people died. In this specific case, they were overwhelmingly elderly women, living alone, and in the upper levels of walk-up apartments.

ELDERLY WOMEN, ALONE, UPPER LEVELS

Consider that the vast majority of those who died in Paris in the 2003 heat wave were "elderly women, living alone, and in the upper levels of walk-up apartments".

In that short phrase, there are three possible risk factors.

First, "elderly". In many, but not all, cases, elderly people tend to have smaller reserves of strength.

Second, "women, living alone". If they collapsed due to dehydration or heat exhaustion, there would be nobody to see their distress and help them. In many cases, their families were on holidays at the seaside.

Third, "upper levels of walk-up apartments". In a heat wave, the upper apartments always get hotter than the apartments underneath. (It would be no consolation that these upper apartments were giving some degree of heat insulation to the apartments below.)

@DoctorKarl **Is grey hair different from white?**

Excessive heat seems to be especially harmful to the very young and the very old, and also to those with chronic diseases and mental illnesses. Related risk factors include being obese, very malnourished, and very unfit.

Another factor is drugs – both legal and illegal.

Dehydration combined with alcohol consumption makes the situation worse.

When the electrical power grid crashes, the loss of air-conditioning in poorly designed houses can be fatal. Another factor in Europe is that the houses are generally designed to keep the heat in, not out.

AUSTRALIAN HEAT WAVES

The current Australian trend (as compared to 50 years ago) is that heat waves arrive earlier in the season, are hotter, and last longer.

Australia-wide, the number of days per year where the maximum temperature is hotter than 35°C has doubled over the last 50 years.

In Melbourne, the heat wave season now starts 17 days earlier. In Perth, the number of heat waves in a calendar year has doubled. In Adelaide, heat waves last two days longer then they used to. In Sydney, the heat wave of February 2011 sent 595 people to hospital emergency departments, and killed 96 people.

In Melbourne in 2014, during the Australian Open, a heat wave struck, bringing not just four consecutive days with temperatures over 41°C, but also the hottest-ever 24-hour period on record. The health risks for both spectators and tennis players were so great that matches in the multi-million dollar competition were postponed after more than 1000 spectators had to be treated for heat stress.

NUMBERS (SORRY) . . .

This rather mathematical statement about the 2003 European heat wave says that in France "the mean summer temperature (June to August) was 3.6°C (or 3.5 Standard Deviations) above the long-term mean". Are 3.5 Standard Deviations (whatever they are) noteworthy?

1) Each hair shaft is made by a "Hair Follicle =HF" 2) HF has "Sub-Factory = SF" that makes/injects dye 3) SF dies → white.

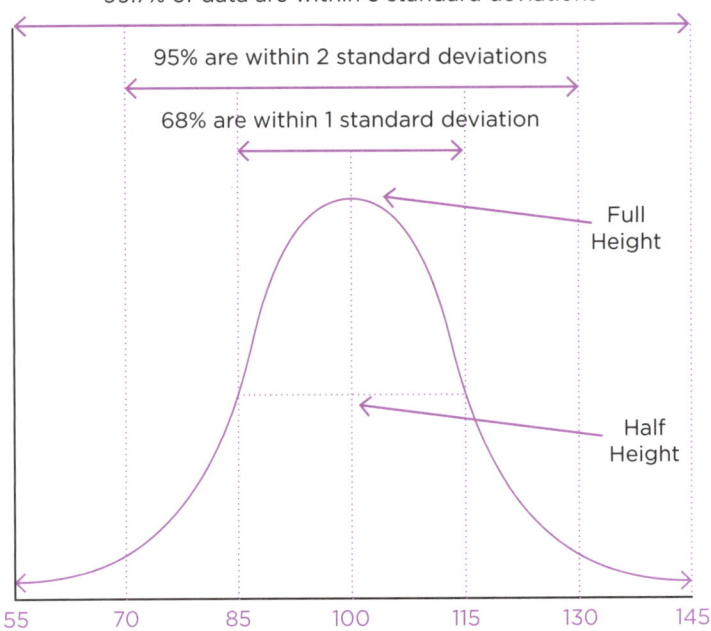

99.7% of data are within 3 standard deviations

95% are within 2 standard deviations

68% are within 1 standard deviation

Full Height

Half Height

55   70   85   100   115   130   145

"Standard Deviation", when married to the famous "Bell Curve", tells you about the shape of the Bell Curve – whether it's low and wide, or tall and skinny.

I have always been amazed by how the Bell Curve applies to so many natural processes. They include the measurement of Intelligence Quotient (IQ) in a population, the diameter of tree trunks in a forest, the heights of waves at sea – and heat waves.

Take the example of IQ. First you measure the IQ of thousands of people. Next the average gets "adjusted" to 100. Then it turns out that the Standard Deviation is 15. In fancy mathematical talk, "the band between the average minus one Standard Deviation, and the average plus one Standard Deviation, will contain about 68 per cent of your population". In plain English, it means that about two thirds of people (68 per cent) have IQs between 85 and 115.

When you spread your net wider, the average plus or minus two Standard Deviations, IQs 70–130, Wikipedia says that will include about 95.5 per cent of your population. For three Standard Deviations,

IQs 55–145, you're capturing about 99.7 per cent. So an event outside three Standard Deviations is quite unlikely – it's approximately 0.3 per cent of instances.

Now you can understand the significance of the abnormally high summer temperatures in France in 2003. It was a very uncommon event. It would happen less than 0.3 per cent of the time – say, once every three centuries.

CLIMATE CHANGE AND HEAT WAVES

In 2013, the then-Australian Prime Minister, Tony Abbott, and his Environmental Minister, Greg Hunt, claimed there was no relationship between Climate Change and extreme bushfires. They were almost certainly incorrect.

Dr Thomas Knutson and colleagues from the Geophysical Fluid Dynamics Laboratory at Princeton University in the USA specifically studied this relationship. Their research examined temperatures over Australia and the Western Tropical Pacific. It showed that climate change almost certainly (with a very high degree of confidence) caused the conditions of extreme heat that Australia experienced in 2013.

MORE HEAT WAVES?

Back in the early 1960s, in any given year, heat waves with temperatures three Standard Deviations above the average covered only about 1 per cent of our planet's land area. By 2010, this had risen to about 5 per cent. By 2020, it's expected to rise to 10 per cent – and for 2040, to 20 per cent. In other words, before the middle of this century, in each calendar year when heat waves do arise, they will cover about one fifth of all the land area on Earth.

So watch it – heat can pack a powerful punch, and can even knock the living daylights out of you.

Just add $$ plus Political Will.

HEAT COSTS AUSTRALIAN ECONOMY

Heat stress costs the Australian economy about US$6.2 billion per year.

You've probably heard of "absenteeism", staying home from work. "Presenteeism" is the phenomenon of turning up at work, but performing less efficiently. With regard to heat stress, you can be affected by heat while at work, or you might come to work exhausted by a poor night's sleep. The

2015 HEAT WAVES

In the mid-June 2015 heat wave in Pakistan, over 1000 people died in Karachi alone.

A fortnight earlier, a separate heat wave in India had just eased. Over 2500 people had died as a direct result of that heat wave.

These death estimates almost certainly significantly underestimate the true toll.

AGRICULTURE AND HEAT WAVES

A commonly heard "claim" is that Global Warming will improve food production by increasing CO_2 levels and prolonging the growing season. The claim seems to be based on the truism that if a little of something is good, then surely more must be better.

This simplistic claim turns out to be incorrect. (After all, you can drink one litre of water and live. But drinking 10 litres all at once will kill you.)

Increasing CO_2 levels can increase the gross amount of vegetation – but that plant matter is nutritionally degraded. Unfortunately, while the plants now have higher carbohydrate levels, they also have lower protein levels, and less magnesium, potassium and the like. The levels of zinc, iron and protein are dropping in wheat and rice. At least two billion people get their iron and zinc from rice. So the food crops

@DoctorKarl Take a solid ridiculously long object and push one end. How long before the other end follows suit?

negative effects include lapses of concentration, more accidents at work, and poor decision-making.

One survey found people were taking 4.4 days off from work per year due to heat stress. Some 70 per cent of people were less productive on at least one day in the previous year.

Heat-Related Presenteeism cost the Australian economy US$932 per worker per year. Absenteeism cost US$845 per worker.

So the total cost of heat stress worked out to about 0.4 per cent of Australia's GDP. But the cost of reducing Australia's carbon emissions to zero by 2050 is less – 0.1 to 0.2 per cent of GDP.

Go figure . . .

---

grown in high CO_2 levels can deliver a double whammy – obesity in wealthy countries, and protein deprivation in poor countries.

Hotter temperatures in general, and heat waves specifically, cause excess evaporation. They also speed up plant growth – but at the wrong times, so this can lead to reductions in crop yields.

In the 2003 Western European heat wave, there were record high daytime and night-time temperatures. This reduced the "leap and grain-filling development of key crops such as maize, fruit trees and vineyards". The heat wave also brought forward the time of crop ripening and maturity by 10–20 days. This change in timing, coupled with reduced soil moisture, led to massively increased water consumption in agriculture.

As a result, in Italy the maize yield dropped by 36 per cent.

In France, temperatures were 3.6C° above normal during the heat wave. Maize and fodder production subsequently dropped by 30 per cent, corn and fruit harvests fell by 25 per cent, while wheat harvests declined by 21 per cent.

In 1972, a heat wave in south-west Russia and Ukraine devastated their cereals harvest. As a result, world grain prices tripled – from US$60 to US$208 per metric tonne.

One obvious region of the world that is susceptible to higher temperatures is where half the world's population live – the tropics and subtropics. Projections are that the population of this region will double to six billion by 2050, while food production is expected to drop by around 30 per cent – thanks to the dryer climate putting stress on agriculture. This is a potentially disastrous mismatch.

Impulse travels through solid/gas/liquid at Speed of Sound (relevant to that object/material/element).

Lesk et al. conducted a study in 2016 to look at the effects of extreme weather disasters on global crop production, from 1964–2007. Overall, droughts and extreme heat reduced cereal production by 9–10 per cent.

Crop failures linked to extreme weather are increasing. Consider crop failures that currently happen once every century. By 2050, they are projected to happen once every 10 years.

On average, weather patterns are moving from the Equator towards the Poles at about 5 kilometres per year, which works out to 50 kilometres per decade, and 500 kilometres per century. This means big changes for agriculture.

We need to quickly develop crops that are both heat-resistant and don't need much water.

Agriculture faces three huge research challenges. It has to increase yields to cope with a growing population, lessen its impact on the environment, and become more resilient to weather that is more extreme.

@DoctorKarl Why does exploding a bomb in water have less impact than exploding in air?

HOW TO DEAL WITH A HEAT WAVE

These seven steps are based on NSW Department of Health guidelines. Basically, you need to drink lots of water, keep yourself cool, take care of friends and family, and most importantly, have a plan to deal with the heat.

1. Drink lots of water, and carry some with you when you leave the house.

2. Avoid drinks that are hot, alcoholic or sugary. (But read "Hot Tea Cools You Down" on page 122.)

3. Plan your day around the heat. Keep indoors between 11 a.m. and 5 p.m., and minimise physical activity.

4. During the day, keep the windows closed for as long as the inside is cooler than the outside. (Yes, the air inside is more stuffy, but it is cooler.) Stop heat from entering the house by shading windows with an awning, shade-cloth or plants. Shut the curtains or blinds to stop direct sunlight entering the house.

5. Check that your air conditioner works properly. If you don't have an air conditioner, and the heat is intolerable, go to an air-conditioned cinema, library or shopping centre.

6. Avoid tight clothing made from synthetic fibres, and wear lightweight, loose-fitting clothing made from natural fibres.

7. When you're outdoors, protect yourself from the sun with both a hat and sunscreen.

Force of explosion has to shift more mass (water about 800x more dense than air).

08 ROADS OF ICE

BACK IN THE YEAR 1557, OVER 28 DAYS IN THE DEPTHS OF WINTER, A TEAM OF MEN MOVED A HUGE ROCK SOME 70 KILOMETRES FROM A QUARRY TO ITS RESTING PLACE IN THE FORBIDDEN CITY IN BEIJING.

Now here's the weird bit. Instead of using spoked wheels, which the Chinese had developed some 3000 years earlier, they chose to drag it.

This rock weighed about 123 tonnes, and was about 9.6 metres long, 3.2 metres wide and about 1.6 metres high.

The heaviest load the Chinese could shift on a wheeled cart was around 87 tonnes. The difficulty with using rollers (such as logs the size of telegraph poles) was getting around corners.

So they put it onto a wooden sledge and dragged it.

If they had simply dragged the sledge across bare ground, they would have needed around 1500 men. If they had laid down a 70-kilometre track of wood, and then dragged the sledge over the wood using water as a lubricant, they would have needed about 350 men.

Now, the average temperature in Beijing in the 15th and 16th centuries in January was around −3.7°C. So the project planners cleverly made

@DoctorKarl Dr Karl, why do you answer stupid questions . . .?

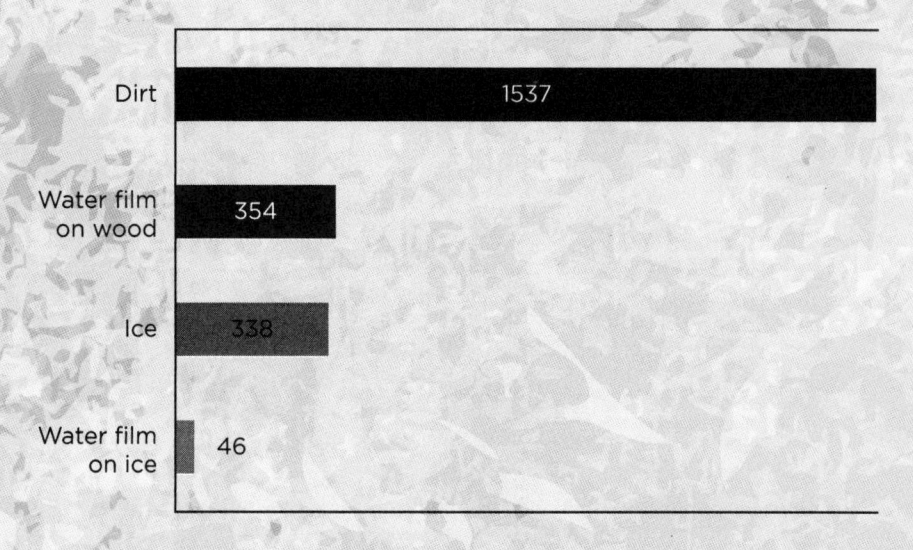

Number of men required to pull slab of rock on various surfaces

Surface	Men
Dirt	1537
Water film on wood	354
Ice	338
Water film on ice	46

a road of ice. If you pour water and let it freeze, the ice will be both flat and hard – easily able to support the weight of the rock. But the friction of wood on ice is only slightly less than wood on wood. You would need only slightly fewer men to keep the stone moving – about 338.

So the Chinese dug wells every half kilometre along the 70-kilometre track, and lubricated the wooden sled with water. Suddenly the friction dropped so much that they needed only 46 men to keep the sledge moving. They could move the sledge at 8 centimetres per second, which worked out to about 290 metres per hour, or 2.3 kilometres each 8-hour day.

Stationary Friction is greater than Moving Friction, so they needed an additional 270 men to get it moving at the beginning of each day. By a nice coincidence, that's also how many they needed to push it up a 10-degree slope.

Their slippery slope led to success, not ruin.

There are no stupid questions . . . they can all lead to enlightenment.

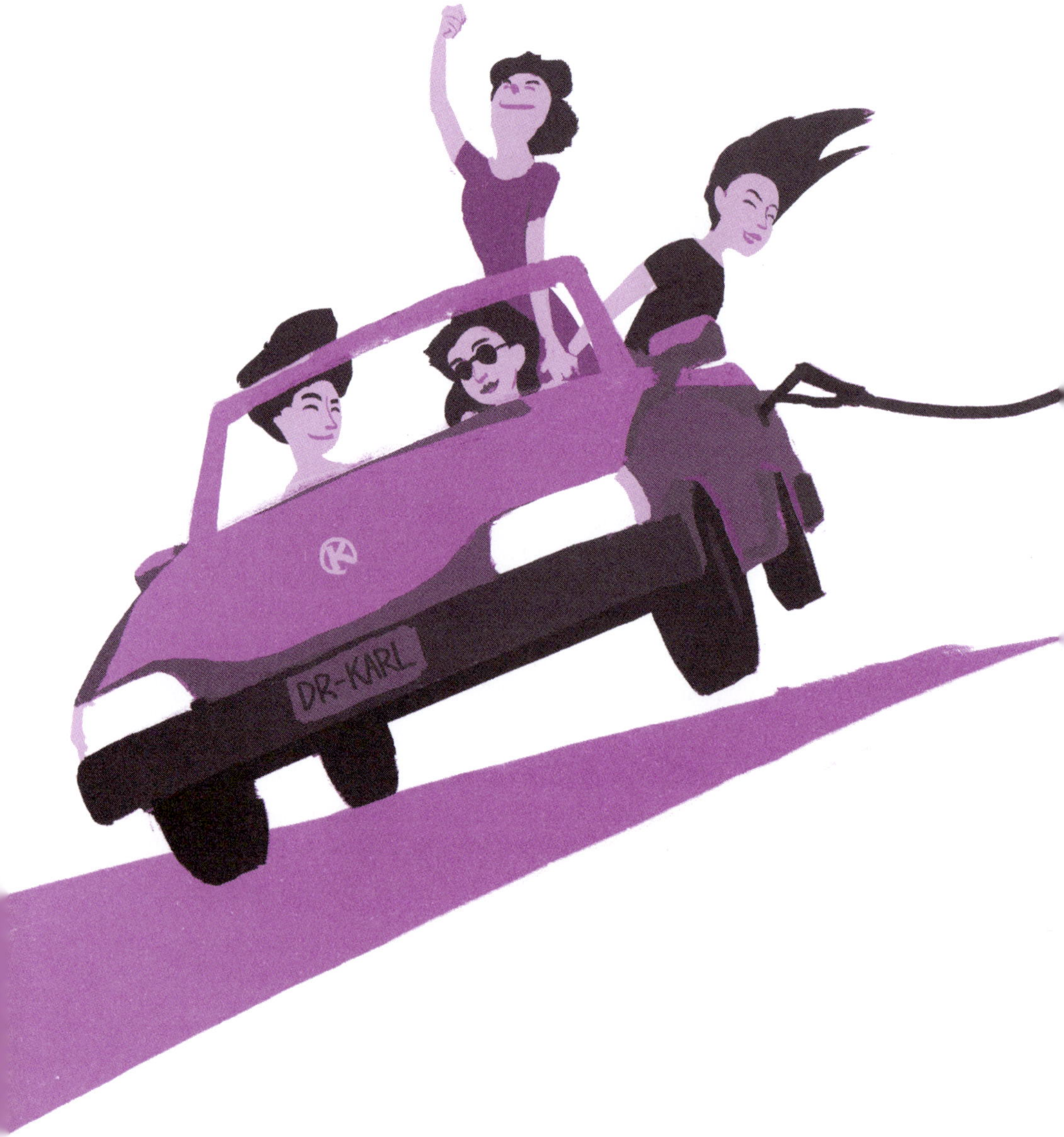
DR-KARL

09

REFUEL CAR WITH ENGINE RUNNING?

IS IT DANGEROUS TO LEAVE YOUR ENGINE RUNNING WHILE YOU REFUEL YOUR CAR? PROBABLY NOT, THE OVERWHELMING MAJORITY OF THE TIME. (BUT IN MOST PLACES, IT IS ILLEGAL.)

Think back to the traditional Fire Triangle that you learned in your first "Fire Safety Officer" course – fuel, an oxidiser (usually the oxygen in the air) and heat. You need all three to start a fire. (Because it needs continuous supply of all three factors, a fire is more of an "event" than a stable "thing".)

FIRE TRIANGLE PART 1: OXYGEN

The air surrounding us is about 20 per cent oxygen. While you are filling your tank with petrol, you are therefore also surrounded by one of the three parts of the Fire Triangle.

FIRE TRIANGLE PART 2: FUEL

You can definitely smell petrol vapour when you refuel. This fuel is the second part of the Fire Triangle. But it usually won't burn.

Petrol vapour burns only in a very narrow range of concentrations – between 2 and 8 per cent of the volume of the air. When the concentration is less than 2 per cent, there's not enough fuel to sustain a flame. And when it's over 8 per cent, there's not enough oxygen available to combine with the petrol vapour – so no burning. The chance of having the right concentration of petrol vapour immediately next to something hot (the third part of the Fire Triangle) is very low.

FIRE TRIANGLE PART 3: HEAT

Yes, there are several potential sources of heat in a car.

The starter motor on the engine draws a lot of current, and can create hot electrical sparks. But it's well and truly buried in the guts of the engine, where the outside air (with or without petrol vapour) has no access.

The spark plugs and their leads have the job of igniting the petrol–air mixture inside your engine. They carry tens of thousands of volts of electricity – but they are normally well insulated. In fact, if the

@DoctorKarl Does every chromosome in any living creature have the same DNA?

installation breaks down, it could ignite petrol vapour (because of electrical sparks along the leads). But you would have difficulty in starting and running the engine, especially in wet weather. So replace any bad spark plug leads.

The inside of the catalytic converter (which cleans up the engine exhaust) can get very hot. The older ones have a safe limit of 750°C, while some newer ones are rated at 900°C. But they cool down quickly, and are buried inside the exhaust system – as close as possible to the engine, and quite a long way from the exhaust pipe opening.

It is also possible, but very unlikely, that there could be random sparks from a faulty relay, or loose battery terminals, or any poorly installed electrical accessory. So be careful if you're putting in that sick new subwoofer yourself.

Sparks of static electricity can theoretically ignite fuel vapour although it's very unlikely. (I discuss this in my 28th book, *Never Mind the Bullocks*.)

LOGICAL INCONSISTENCY?

Like the law-abiding citizen you are, you pull up at the petrol pump, turn off the engine and then start refuelling. Meanwhile, other motorists drive past you *with their engines running*! Nobody expects you to switch off your car's engine while still on the road, well away from the petrol, and then gruntingly push the car to the petrol pump. That's because there's no regulation legislating that running engines are not allowed near somebody refuelling. The only rule is that *your* engine can't be running. Does that make sense?

However, it is definitely dangerous to leave the engine running with children inside the car. They might want to copy their parents and shift the gearbox into drive. Or, while you're inside paying for petrol, somebody could hop into the driver's seat and drive away.

I guess that it's better to be safe than sorry, and not be "fuelish"...

---

1) total DNA/cell = sum of DNA in all 46 chromosomes of cell.
2) Chromosome 1 has different DNA from Chromosome 46.
3) No. Mutations with time.

RISKY BUSINESS

In the United Arab Emirates, local regulations are that the engine should be switched off while refuelling. Most locals ignore that. It gets so hot in summer that many drivers leave the engines running with the air-conditioning on full, for the comfort of the passengers. In most cases, the cars are refuelled by petrol station staff. The drivers and passengers often smoke cigarettes, with the windows slightly down to vent the smoke.

A car did catch fire at a petrol station in Dubai in 2015 and there was a similar fire in the UAE a couple of weeks later. The fire was blamed on "a poorly maintained car".

Refuel Car with Engine Running? **< 47**

In general, no. Immune System sets off reactions against "foreign" chemicals.

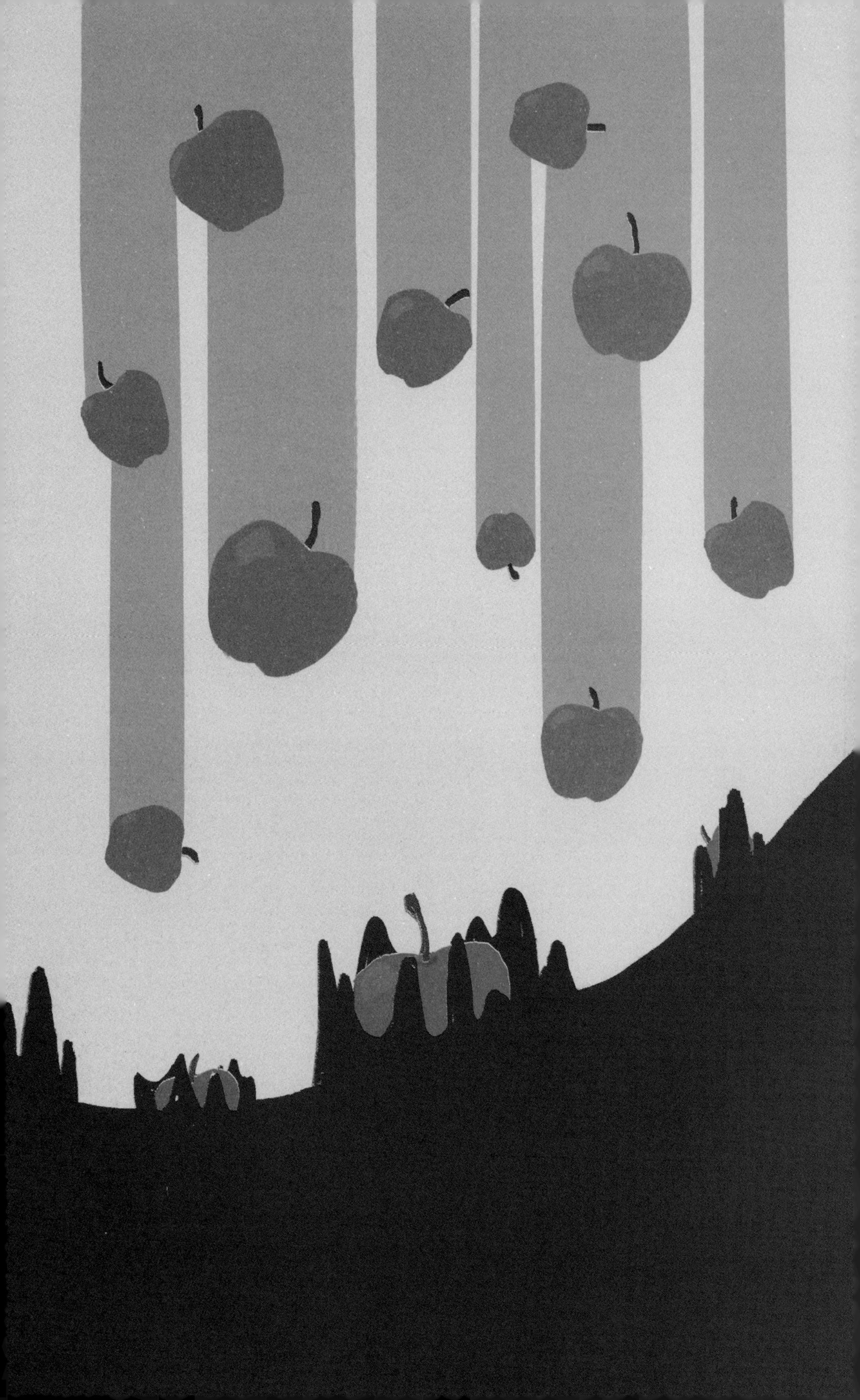

10

GRAVITATIONAL WAVES

ABOUT 1.4 BILLION YEARS AGO, AND ABOUT 1.4 BILLION LIGHT YEARS AWAY, TWO BLACK HOLES RAN INTO EACH OTHER AT HALF THE SPEED OF LIGHT.

In that brief instant they generated more than 50 times the combined power output of the billion trillion stars in the entire known Universe. That enormous power spread out in all directions as Gravitational

COINCIDENCE?

On 14 September 2015, at 3.58 p.m. Australian Eastern Standard Time (AEST), Tony Abbott was deposed as Prime Minister of Australia, and Malcolm Turnbull was elected Liberal Leader and the new Prime Minister.

At 7.50 p.m. AEST (232 minutes later) scientists detected a Gravitational Wave for the very first time as it swept through the Fabric of Space–Time (and our planet).

At that time, all living creatures on Earth were just single-celled critters – and lived in the oceans. It took another half billion or so years for life to evolve into multi-celled creatures, and another half billion or so for it to leave the oceans and crawl onto land.

After a long time, and a long journey, those Gravitational Waves first arrived at our planet on 14 September 2015, at 0951 Universal Time (previously known as Greenwich Mean Time). At the Speed of Light, these strange waves of energy stormed at our Earth from roughly the direction of the Magellanic Clouds. (The two Magellanic Clouds are mini-galaxies, visible on a really dark night, just near the Milky Way.)

They rippled straight through our entire planet. In response, our planet actually changed its shape eight times in a very short window of time – about two fifths of a second. With each of those eight impulses, the entire planet shrank in one direction, and simultaneously expanded in the other direction.

Each impulse was bigger than the one before it. The eighth one was the biggest. And then, after just two fifths of a second, the Gravitational Waves were gone – heading outward for their next target.

In the words of *Doctor Who*, it was classic "wibbly-wobbly timey-wimey stuff". We had just been hit by a ripple in the Fabric of Space–Time. When I heard of this discovery, tingles ran up and down my body for about a minute.

@DoctorKarl Does heavy water (D_2O) weigh more than regular water (H_2O)?

They wrote a paper on their astonishing discovery.

And who were the first three authors of the Gravitational Wave paper?

Abbott, Abbott and Abbott.

Coincidence? You decide . . .

GRAVITY 101

So what is "Gravity"?

(Here's the spoiler. Gravity is a "dent" in Space–Time.)

The Nature of Gravity is a very deep question. Newton pondered and got us part of the way, but it took Albert Einstein to get us to an answer.

Let's start with the basics.

We can see that the world around us has three Space dimensions. You can travel back and forth (the first dimension), left and right (the second dimension) and you can travel up and down (the third dimension).

But there's a fourth dimension as well – Time, which normally travels forward at one second every second.

Put all these four dimensions together, and they make the very literally named Space–Time Continuum (or Fabric) of the Universe. Don't think of Space and Time as separate – they are inseparably mingled.

Just as in the real world, you might meet somebody at the corner of X and Y Streets (two dimensions so far), on the second floor (there's the third dimension) at midday (and that was the Time dimension).

So now it's time for a deep and fundamental insight – and in just three words. (Here it comes.)

Gravity is Geometry.

Imagine a perfectly flat trampoline. The trampoline is our model of the Fabric of Space–Time. When it's flat, that "flatness" is equivalent

Ice cubes made of heavy water will sink. So 1 litre of heavy water has more mass than 1 litre of regular water.

to no Gravity at all. Send an object (anything from a marble to a comet, or even bigger) across it. It will just go in a straight line. There's no dent (Gravity) to swing it off course.

Now put a bowling ball on that trampoline – it makes a dent. Let's pretend that our Sun is that bowling ball. So our Sun makes a local distortion in the fabric of Space–Time. That "local distortion" or dent is a "gravitational field". (Yes, "Gravity" is "Geometry".)

Along comes a totally unsuspecting comet. If there were no Sun, and no dent in the Fabric of Space–Time, that comet would continue in a straight line. But there is a dent in the shape of Space–Time, and so the path of the comet is deflected. Depending on its original path and speed, it might smash into the Sun, or it might come very close to the Sun and loop around it, or it might just very slightly have its course changed.

So Gravity is Geometry – it's just a local distortion (or dent) in the Fabric of Space–Time.

GRAVITY IS DIFFERENT

Gravity is one of the Four Forces that are known to exist. We use these Four Forces to explain most things in the Universe. (At least, "most things" in terms of Physics – we still can't explain dreams, or why people voluntarily look at cat videos.)

Those other three Forces are the Electromagnetic Force (radio waves, TV waves, atoms pushed away from each other, etc.), the Strong Nuclear Force (holds the centre of the atom together, etc.), and the Weak Nuclear Force (involved in some radioactivity and how atoms decay, etc.).

But here's the first big difference.

Think of a theatre. The Space–Time Continuum is the underlying "stage". Gravity (one of the Four Forces) is actually a distortion in that stage (like a hill or a hole). The other three Forces are "players" that perform on top of the "stage". The other three Forces can't affect Gravity – but Gravity can affect them. Gravity (the hill or hole) affects their path.

The Electromagnetic Force, the Strong Nuclear Force and the Weak

@DoctorKarl Does the act of eating celery burn up more calories than you get from the energy in the celery?

Nuclear Force all do their stuff on top of the Fabric of Space–Time. They don't affect the Space–Time Fabric. There is no Electromagnetic or Nuclear "Process" that we can do that will affect Gravity.

The second big difference between Gravity and the other three Forces is that Gravity is the weakest of the Four Forces. When I stand up, I use my muscles. They can overcome the gravitational pull of the entire Earth. Compared to the other three Forces, Gravity is very weak.

GRAVITATIONAL WAVES 101

So now we are ready for Gravitational Waves.

(Here's the spoiler. When dents in the Fabric of Space–Time accelerate, they give off Gravitational Waves. The dents have to "accelerate", not just move at a constant velocity. These Gravitational Waves then spread quickly outwards in all directions – at the Speed of Light.)

Let me start with two examples.

When you wave a stick back and forth through water, you get water waves. With your eyes, you can see a cork bob up and down as the waves travel past it. (This is actually not a very good example – because you get waves in the water when you move the stick at a constant velocity, not just when it's accelerating. But it's a start.)

When you move electrons back and forth, you get electromagnetic waves. You can detect the changing electrical and magnetic fields with a radio, TV, or appropriate measuring device.

If I move my hand back and forth – making it speed up and then slow down, i.e. accelerate – it will emit Gravitational Waves. My hand has its own tiny amount of Gravity. It distorts the Space–Time around it. And when I accelerate my hand, that distortion moves with it. Furthermore, ripples of distorted Space–Time move outwards in all directions from my hand at the Speed of Light.

But Gravity is incredibly weak. The Gravity of the entire Earth can't pull a fridge magnet away from the metal door of your fridge.

So when I accelerate my hand, because Gravity is so weak, the

100 gm celery → 32 kJ (digestible carbs) + 16 kJ (indigestible carbs, gut bacteria). But Thermic Tax = 4 kJ → 44 kJ/100gm celery.

amount of power in the Gravitational Waves from my accelerating hand is microscopic.

Even if you go for something a lot bigger, like the Earth orbiting the Sun, Gravitational Waves are still tiny. (By the way, the Earth's orbit is not a straight line. It's curved, and so the Earth is always accelerating – to change direction and keep following a curved path. If a mass is accelerating, it gives off Gravitational Waves.) Because Gravity is so weak, the power in the Gravitational Waves emitted as the Earth orbits the Sun is tiny – only about 200 Watts. That's tiny – a decent car engine can generate a thousand times more power.

GRAVITATIONAL WAVES (ADVANCED)

About a century ago, Albert Einstein came up with his Theory of General Relativity, which is really a Theory of Gravity. One of Einstein's many great insights was to treat both Space and Time as real things. He was the first to see Gravity as a distortion of Space–Time. Einstein said that any accelerating mass would give off Gravitational Waves, which would then ripple through the background fabric of Space–Time at the Speed of Light.

A Gravitational Wave is a moving distortion in the fabric of Space–Time – it travels *through* Space–Time, not on top of it. (I'm just emphasising, by repetition, this fundamental point – sorry.)

So Gravitational Waves are ripples in the Fabric of Space–Time. They are emitted from accelerating masses. They travel outwards in all directions, at the Speed of Light.

They are different from Electromagnetic Waves (such as light, microwaves, X-rays and so on), which travel *on top of* this Space–Time Fabric of the Universe.

Gravitational Waves are a travelling distortion of the actual Space–Time Continuum of the Universe. Gravitational Waves are the actual rippling and changing of the shape of Space–Time itself. (Yes, they still are 100 per cent genuine "wibbly-wobbly timey-wimey stuff".)

DECAYING ORBITS

Think about objects in orbit. They are continually changing direction, so they are accelerating – and so they are emitting Gravitational Waves. These Gravitational Waves carry power away from the "system" – and so the orbits will decay.

Let's start small.

The Earth orbits the Sun. So it radiates about 200 Watts in Gravitational Waves. The "Energy Equations" have to balance. So every year, the Earth's orbit shrinks – but only by about one three-hundredth of the diameter of a hydrogen atom each year, which is about a third of a trillionth of a metre.

If you look at all the planets in the solar system, they radiate away a total of about 5000 Watts.

Let's move up to a neutron star and a pulsar in orbit around each other. This pair of orbiting stars is radiating about seven trillion trillion watts – about 2 per cent of the Sun's output. They are getting closer by about 3.5 metres per year.

Let's ramp it up a bit more. Consider two neutron stars, each of about two Solar Masses, orbiting each other at about 189,000 kilometres. They would orbit each other in about 20 minutes, and would radiate Gravitational Waves about 100 times the power of our Sun – a total of 10,000 trillion Watts. The neutron stars would get closer by about 116 metres each year – and would smash into each other in about 414,000 years.

The proposed Evolved Laser Interferometer Space Antenna (eLISA), a trio of satellites set to be launched by around 2030, could detect these Gravitational Waves from neutron stars orbiting each other.

So many discoveries are yet to be made with better telescopes . . .

Yes. Cold weather → peripheral vascular shutdown (to preserve heat) → ↑blood to kidneys →↑Glomerular Filtration Rate →↑urine.

BIG COLLISION

Back on 14 September 2015, we humans detected our first Gravitational Waves. Gravity is a very weak force, so only truly extreme events will produce detectable waves (at least, with our current technology).

A pair of black holes colliding with each other turns out to be that kind of cataclysmic event. These two black holes were much heavier than our Sun. One was 29 times heavier. (The Astronomers call the mass of our Sun "one Solar Mass". So this black hole had 29 Solar Masses.) The other black hole had 36 Solar Masses.

We were able to pick up the last two fifths of a second of their death spiral. But detection was possible only when our technology finally got good enough to detect Gravitational Waves – and if the event was Big.

According to the mathematics given to us by Albert Einstein in his Theory of General Relativity back in 1915, these black holes had been emitting energy before their collision, in the form of Gravitational Waves. Because they were losing this energy, their orbits around each other were shrinking. Einstein's maths told us that as they got closer, they would lose energy much more quickly.

It was inevitable – they were definitely going to slam into each other. The collision was going to be big. And our detectors here on Earth had, only days before, been upgraded to be just sensitive enough to pick up the Gravitational Wave energy that they pumped out in their last eight orbits before their collision. (What lucky timing!)

Over those last eight orbits, they accelerated from circling each other 17 times per second right up to 75 times per second – and then they collided.

At that very last instant, immediately before that cataclysmic collision, they were travelling at about half the Speed of Light – roughly 150,000 kilometres per second. And remember, these were not inconsequential objects, like tiny sewing machine parts. No, these were black holes, with masses 29 and 36 times that of our Sun!

With that gigantic impact, they somehow merged into just one black hole, weighing in at 62 Solar Masses. But if you add the mass of the initial black holes at 29 and 36 Solar Masses, you get 65 Solar Masses, not 62. You can see that three times the mass of our Sun went missing.

So where did it go?

That mass got turned into pure energy – lots and lots of it. And then that energy got turned into Gravitational Waves.

POWER OF THREE SUNS

Let's try and get a grasp of what it means to turn three Suns' worth of matter into pure energy.

Let's start small, and work our way up.

Each second, our Sun burns about 620 million tonnes of hydrogen to make energy. Our Sun is halfway through its life, but over its entire existence of about 10 billion years, it will burn up less than one thousandth of its mass.

Now consider what these two colliding black holes did. What they turned into energy was not just one thousandth the mass of our Sun – but three times its total mass! And they did this, not over 10 billion years, but in less than one fifth of a second.

In that instant, they temporarily generated over 50 times the power output of all the billion trillion stars in our entire Universe put together! (In more precise numbers, the power was around 3.6×10^{49} Watts, or 36 trillion trillion trillion trillion Watts.)

If you were close to that collision, when all that power let loose, everything would have been very messy. Right at the location where the two black holes joined into a single black hole, a 2-metre-tall person would have been stretched to 4 metres, and then, within a thousandth of second, been shrunk to 1 metre. But we're 1.4 billion light years away – and so, luckily, that power was diluted over a huge volume of space, about a billion cubic light years.

And as the biggest of those Gravitational Waves rippled through our planet, it changed our planet's size and shape. Earth is about 12,750 kilometres across. It shrank, and then expanded, by a very tiny amount

Excessive blue light (from TV, laptops, etc.) → interfere with bipolar cells in retina → melatonin from pineal gland → poor sleep.

– roughly two-and-a-half times the diameter of a proton, which works out to two-and-a-half times a thousandth of a millionth of a millionth of a metre.

You can see that there's no way our ordinary human senses could have detected this tiny change. (Maybe Obi-Wan Kenobi from *Star Wars* could have "felt" this tiny change in the Force – but he was a Jedi Master.)

However, the astronomers did notice, thanks to a pair of revolutionary "telescopes" each called the Laser Interferometer Gravitational Wave Observatory (LIGO). But to turn Einstein's theories into LIGO took a long time – about a century, in fact.

DO GRAVITATIONAL WAVES EXIST?

Why did it take a century? Well, there were a few decades of speculation, followed by four decades of doubt as to whether Gravitational Waves could be detected, and finally a quarter century of building successively more sensitive machinery until we actually measured them.

Back in 1916, Albert Einstein formulated Gravitational Waves as a consequence of his theory of General Relativity. But then he dived deeper into the mathematics, and as a result he thought Gravitational Waves might not exist. And then for the next two decades he worried back and forth about it, changing his mind several times. Finally, he decided they should exist.

But could we ever detect them?

By the mid-1950s, there was still doubt as to whether they could be theoretically detected by any instruments we could ever build. Finally, Richard Feynman (my hero) and Hermann Bondi developed the so-called "Sticky Bead" Thought Experiment. They showed that we could (at least in theory) detect Gravitational Waves.

Here's the simple version. Start with some beads threaded onto a sticky rod. Imagine that a Gravitational Wave comes rippling along and accelerates the beads. These beads would move and (thanks to friction) transfer some heat to the rod. This is "proof" that Gravitational Waves must carry energy, and that, theoretically, they are detectable. But in the 1950s, the existing technology was too insensitive to detect these Gravitational Waves – millions of times too insensitive.

Two decades later, in the mid-1970s, two important events happened.

INDIRECT DETECTION OF GRAVITATIONAL WAVES

About a decade earlier, in 1967, a strange new type of star had been discovered.

It was a "neutron star". It's a star that (after a few billion years of burning) had shrunk down to a ball of neutrons. A typical neutron star is about 20 kilometres across, with a mass of up to about two and a half times that of our Sun. They are incredibly dense – a matchbox of their substance would weigh about 13 million tonnes.

A "pulsar" is a rotating neutron star, with a powerful magnetic field, which emits regular pulses of energy – hence the name "pulsar".

In the mid-1970s, two physicists, Joseph Taylor and Russell Hulse, used the largest radio telescope in the world to find 40 pulsars. One "object" they discovered was actually two objects – a pulsar and a neutron star orbiting each other every eight hours. They whizzed around each other at an average speed around 300 kilometres per second, and came within about 700,000 kilometres of each other at their closest approach. It turned out that their orbits are shrinking at about 3.5 metres per year. They are expected to collide within 300 million years.

Why is the orbit shrinking?

Einstein's theories (again) had the answer.

The orbit was shrinking because the neutron stars were emitting energy in the form of Gravitational Waves. The measured rate of decay of the orbit agreed with Einstein's General Relativity predictions to within 0.2 per cent. The power emitted was about 2 per cent of our Sun's total power output.

Babies have only cartilage at "kneecap = patella", which later ossifies into bone.

But this test did not actually detect Gravitational Waves directly. It just indirectly detected the result of Gravitational Waves being emitted, which was the orbit shrinking. But at least we had measured their effect. (Taylor and Hulse won a Physics Nobel Prize for this.)

EINSTEIN ROCKS

In 1916, Albert Einstein predicted that Gravitational Waves should exist. In 1917, he laid down the theoretical framework for lasers.

One century later, we used lasers to detect Gravitational Waves.

DIRECT DETECTION OF GRAVITATIONAL WAVES

The second thing that happened in the mid-1970s was that two physicists, Kip Thorne and Rainer Weiss, ended up sharing a hotel room at a conference. They spent the whole night doing what Gravitational Physicists do. They excitedly talked about Gravitational Waves and the best way to actually detect them. They then teamed up with a gifted experimentalist, Ronald Drever.

The team settled on a novel detection method. Split a laser beam into two separate beams. Then run the laser beams through two pipes at right angles to each other. These pipes are four kilometres long.

At the end of the long pipes are mirrors that bounce the laser beams back to where they first split. Finally, recombine the beams and look at their interference pattern. (You can see interference patterns in rivers, when the bow waves from passing boats "interfere" with each other.) Once the system is up and running, and the Space–Time Continuum is calm, then nothing should happen to the interference pattern. Then a Gravitational Wave comes rippling through our solar system. It expands and shrinks the Fabric of Space–Time itself. The arms should change in length, and the interference pattern from the two laser beams should change. Bingo! You've just picked up Gravitational Waves.

The current sensitivity of LIGO is such that it can detect a change in length of one of the four-kilometre-long arms of one ten-thousandth of the diameter of a proton. The measured change in the first event of 2015 was roughly ten times larger – one thousandth of the diameter of a proton.

Normally . . .
- Mirrors
- Beam splitter
- Light detector
- Laser

When a Gravitational Wave passes . . .
- Detectable signal

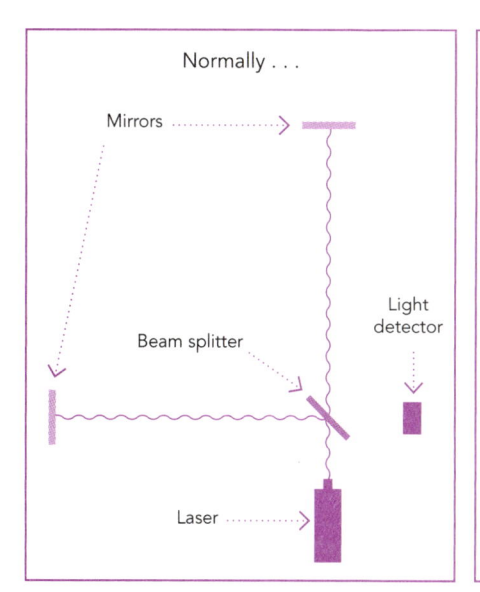

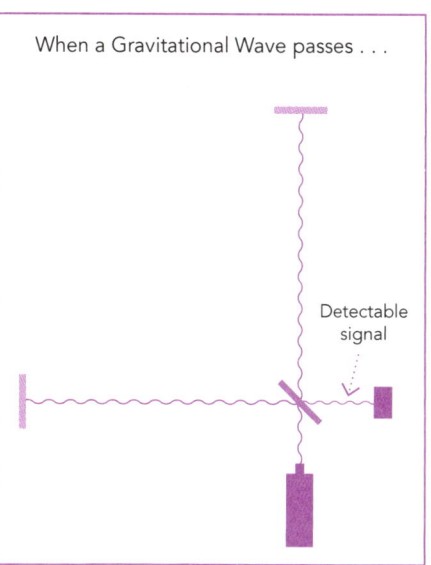

BUILDING LIGO

Of course, building a device capable of doing that kind of measurement wasn't easy. With typical honesty, the scientists said that the detection of Gravitational Waves was "possible" with the first version of LIGO, but "probable" with the later Advanced version. That's exactly how it panned out.

It took decades to get the money – and then decades to build LIGO once they had the cash.

Construction of LIGO began in the mid-1990s. After starting up in 2002, it didn't find any Gravitational Waves. So it was shut down in

Sure. Equator spins ~1,600 kph, Pole ~0 kph, and in between is pro rata. Pick your latitude to circumnavigate Earth.

2010, rebuilt to three times higher sensitivity by 2015, and started up again. Within a few days of starting up, they measured that first detectable Gravitational Wave as it rippled through us.

There are actually two identical LIGOs, one in the state of Washington in the north-west of the USA, the other in Louisiana in the south-east. By having two separate observatories, and recording the time when a Gravitational Wave hits them, you can get a rough idea of where in the sky it came from. As we build more Gravitational Wave detectors, we can more accurately localise the source of the event.

In each of the LIGO facilities, laser beams had to run through the largest volume of vacuum ever constructed – some 10,000 cubic metres of "nothing". The laser beams bounced off mirrors at the ends of the four-kilometre-long tunnels – and these mirrors had to be isolated from vibrations. By using noise-damping suspensions and active feedback systems, hanging the mirrors on glass threads, and lots more, the Louisiana LIGO could tune out not just a nearby door slamming, but the impact of waves in the Gulf of Mexico hitting the beach a hundred kilometres away.

As a result scientists could detect some of the tiniest changes in distance ever in the history of the human race. As an example, if they were measuring the distance to the nearest star (about four light years away), they could see a change of the thickness of a human hair!

The technical skills needed were absolutely mind-blowing – it was Science so advanced that it bordered on Art.

THIRD ASTRONOMY

Now here's something surprising. By detecting Gravitational Waves, we have invented a brand-new branch of Astronomy – the third one so far.

The first branch is Electromagnetic Astronomy. It looks at Electromagnetic Waves that travel on top of the Space–Time Fabric of the Universe. It's been running for millennia, beginning with naked-eye astronomy in the Ancient World (practised by the Egyptians, Mesopotamians, Chinese, and so on).

It really took off about four centuries ago, with Galileo and the telescope – but using visible light only. Centuries later, we expanded into other frequencies in the electromagnetic spectrum, such as Radio Waves in the 1930s, and more recently, Infra-Red, Ultra-Violet, Gamma Ray, X-ray, microwave and so on. All these other frequencies are still part of the electromagnetic spectrum.

The second branch is Neutrino Astronomy. Neutrinos were discovered only in 1956. They can be created by certain nuclear reactions, such as those that happen in the Sun. Every second, about 65 billion solar neutrinos pass through each square centimetre of your body. But thanks to their tiny mass and other factors, they almost never interact with your body – they just pass through.

So they are very hard to detect. To have a 50 per cent chance of catching a neutrino, you need a few light years' worth of that very dense metal, lead. To put that in perspective, you need only a few millimetres of lead to stop X-rays.

As an example of how hard neutrinos are to catch, consider what happened back in 1987, when a nearby star exploded into a supernova. It emitted trillions upon trillions of neutrinos. But down here on planet Earth, we detected just two dozen of them with our three early neutrino telescopes.

Today, we have more advanced neutrino telescopes, such as the IceCube Neutrino Telescope down at the South Pole. It should be joined by another two advanced neutrino telescopes by 2017 – and more discoveries will be made. But, just like Electromagnetic Radiation, neutrinos travel across (or on top of) the Space–Time Fabric of the Universe.

This third, brand-new Gravitational Wave Astronomy is very different. It doesn't look at stuff travelling on top of the Space–Time continuum. No, it looks at actual travelling distortions in the Fabric of Space–Time itself. But the two current LIGO observatories are sensitive only to frequencies roughly between 10 cycles per second and a few hundred cycles per second. With further improvements in sensitivity,

**Each new answer gives two new questions.
We'll never know everything.**

they could pick up rotating neutron stars, or exploding supernovas.

But there's a whole range of frequencies that LIGO is not tuned to (in the same way that an optical telescope does not pick up X-rays). With different Gravitational Wave telescopes, we could pick up compact objects captured by supermassive black holes in the centres of galaxies, or a pair of supermassive black holes orbiting each other. The space-borne eLISA Gravitational Wave satellites are scheduled to be running in around 2030. They should be able to find Gravitational Waves with frequencies less than 1 Hertz – corresponding to black holes with masses of millions of Solar Masses.

WHERE TO?

The braininess of Albert Einstein was truly impressive. Time after time, he went out on an intellectual limb. Not only did he lay down theoretical frameworks that gave us useful gadgets such as the laser, GPS and solar panels, he also made many predictions that were later proven correct – the most recent one being Gravitational Waves.

Now I'm going way out on a limb here. We have just learnt how to detect Gravitational Waves. Maybe somebody living today will give us the framework for being able to *make* Gravitational Waves – and perhaps the Anti-Gravity dream might finally get off the ground . . .

Gravitational Waves **< 65**

**Vast majority of chewing gum leaves body via regular pathway, on time.
Need **huge** quantities to upset timing.**

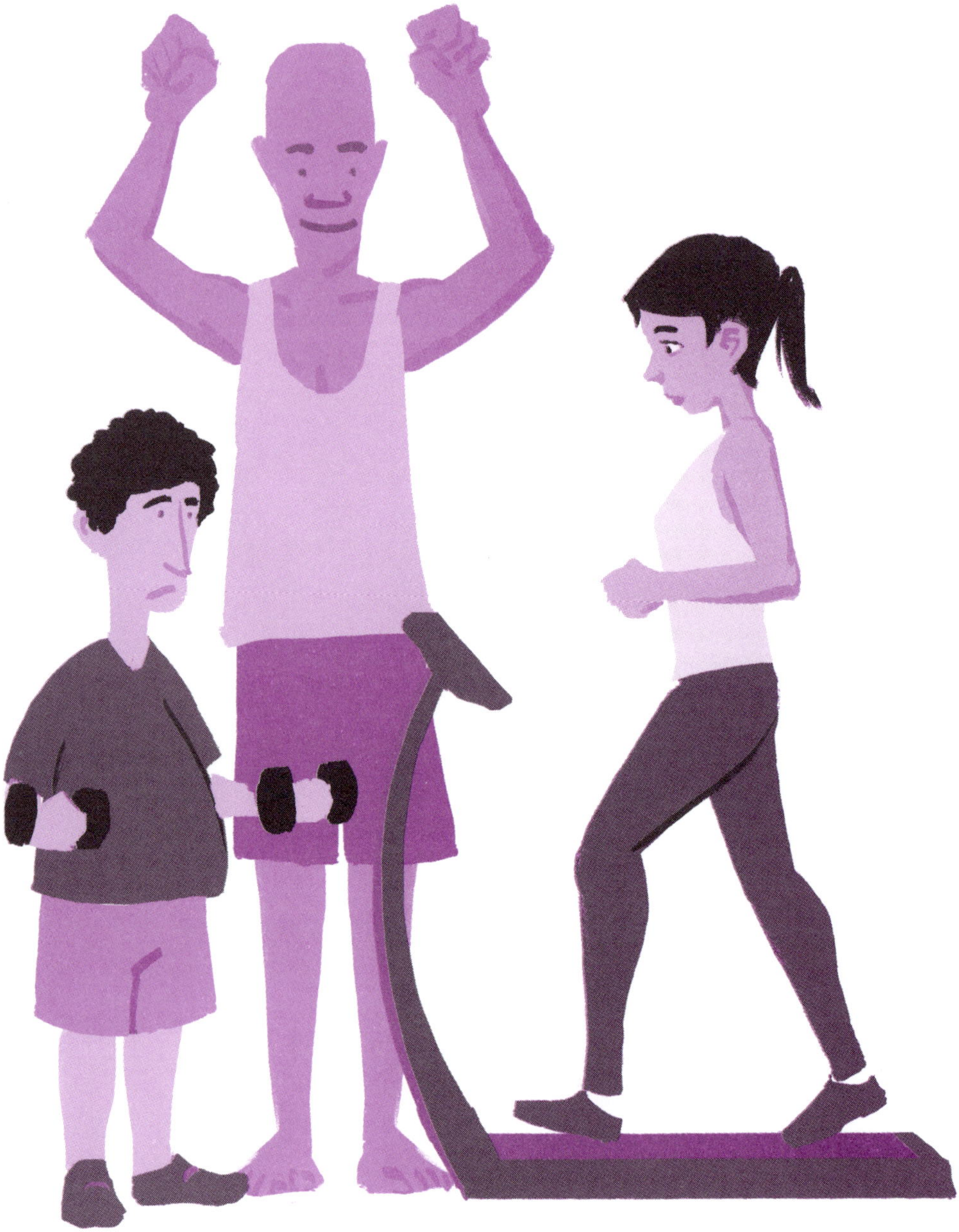

11

THE HEIGHT OF GOOD HEALTH

YOU'VE GOT TO LOVE HISTORY – IN ITS OWN WAY, IT'S THE "ULTIMATE SCIENCE FICTION". HISTORY SHOWS US THE PAST IN MANY WAYS.

Let me use the literal Measure of a Man to see where we've been, and where we're going.

Welcome to Auxology – the study of height. Height is a general marker for many aspects of biological wellbeing, including life expectancy.

So riddle me this little mystery. For about two centuries, the Americans were among the tallest people on Earth. Way back then, the Dutch were down at the short end. What changed so that today the Dutch are the tallest people on Earth, and the Americans are not?

It seems the countries that provide their citizens with Cradle-to-Grave Health Care have the tallest citizens.

HEIGHT 101

We think that height is (like many other human characteristics) due to a mix of nature and nurture. The current estimate is that it's about 80 per cent genetic, and 20 per cent environmental.

A 2014 study examined the DNA of about a quarter of a million people. It appears that about 700 different genes in about 400 different locations help control height. These genes seem to operate in three body zones – the legs, the spine and head, and finally, the overall body.

We have three major growing periods after we're born. There's a huge spurt when we are babies and infants, a little one around age six or seven, and another big spurt during puberty and adolescence.

But environmental factors can be very powerful.

Obviously, diet and nutrition are important to reaching your full potential height. Emotional factors are also important. If a well-fed child is abused or neglected, they will often be stunted in height. For example, it's well documented that children would slow down their growth at boarding school, and then catch up when on holidays with their parents.

Education is another environmental factor that helps you grow taller – and also should make your life better. For example, people who have more education tend to need social services less.

Another factor leading to increased height is the trend to smaller families. This means that each child gets more of their family's time and resources.

@DoctorKarl **How do you measure fizziness in fizzy water?**

HISTORY OF HEIGHT

The timeline of history shows us some quite extreme height differences.

Around the time of American Independence, in the late 1770s, American-born soldiers were 5–7 centimetres taller than their English equivalents. In early 19th century England, the difference in height between upper- and lower-class youth was an incredible 22 centimetres.

Around the mid-1930s, the Americans dramatically slowed down their previously regular height increase per decade. They didn't recover it until the late 1970s – but then, with two oddities. First, their increase in height resumed, but at a much slower rate. Second, it applied only to Caucasians – the adult height of African-American people stood still. In fact, if you specifically look at African-American women, it appears that their average height has dropped by about 1–2 centimetres.

We don't fully understand the American slowdown in height, relative to the Europeans – but we have identified some factors.

CAUSES OF HEIGHT IMPROVEMENT

The first and major factor was generally improving nutrition and health of the European population. This was associated with a simultaneous lowering of disease in the community.

From around 1850 to 1900, there was a decrease in disease carriers in the cities, such as horses (and horse poo) in the street, and pigs and chickens in the backyard. Industrial pollution began to drop. In housing, sanitary conditions began to improve, and there were increasing supplies of fresh food to the cities, especially milk.

But most importantly, around that time, there began the professionalisation of Health Services and Public Health Programming – which lead to general improvement in health and nutrition. Government-sponsored health services began in Germany and Austria in the 1880s, Belgium, Denmark, France and Sweden in the 1890s, and the UK and Ireland in 1911.

Mmm. 1) Add fine powder to each, see how high they fizz (nucleation centres). 2) Shake bottles, fill balloons on neck?

AMERICAN FOOD

Compare current European and American children's diets.

American children eat more meals outside the home, and this food is higher in fat and energy. Unfortunately, this food is lower in essential micronutrients.

The increasing income of families was another factor, allowing better and more food. For example, in the UK around 1900, about 60 per cent of working class income was spent on food. Dieticians have identified some 50 nutrients essential for health and growth. Missing out on any of these subtly nibbles away at your overall health and ultimate height.

TALLEST TO FATTEST

Now remember those lanky Dutch, the current tallest people on Earth? At the end of World War II, the average Dutch male was 1.70 metres tall – about 7–8 centimetres shorter than his American equivalent. (Today, the average Dutch male is 1.84 metres tall.)

One factor unique to the Netherlands is the extensive, free system supporting Dutch mothers from all social classes. All mothers can get expert advice on infant feeding, child nutrition and hygiene from a nurse, general practitioner and a specialist paediatrician.

But the USA has the most expensive health care system on Earth, which is unfortunately combined with a weak social safety net. According to John Komlos and Marieluise Baur, from the Department of Economics at the University of Munich, there is "greater social inequality, an inferior health system, and fewer social safety nets in the United States than in Western and Northern Europe, in spite of higher per capita income".

This means that full and complete medical care has been effectively

@DoctorKarl Are you born with a set number of heartbeats to last your whole life?

Another surprising fact is that each year, 25 million Americans get emergency food rations. Yet, at the same time, many Americans are obese.

denied to both the children and adults in a large proportion of the US population. For example, half of unwell adults in the USA did not see a doctor when sick, or did not get recommended treatment, or did not fill a prescription because of cost. Wealth in America is not fairly distributed across its population. (I wrote about this in "The 1 Per Cent and the 20 Per Cent – Economics and Physics", in my 34th book, *Game of Knowns*.)

So while the USA might be at its height as a military and economic Superpower, that doesn't translate to its citizens standing taller than the rest of the world.

HEIGHT AND INCOME

It seems that the Universe is unfair. Taller people earn more than shorter people, men earn more than women – and the effect is greater in the USA and UK than in Australia.

Consider a 10 centimetres increase in height. That's about 4 inches, or the distance across your clenched fist.

Women first. In the UK and USA, they earn an extra 5–8 per cent for each 10 centimetres of extra height. But in Australia, they get just 2 per cent more.

Now men. In the UK and USA, they'll get 4–10 per cent for each extra 10 centimetres. But in Australia, it's just 3 per cent.

Just to be a bit more unfair, on average, "short people report worse physical and mental health than people of normal height".

Not quite. But if you do cardiovascular exercise, heart rate slows down + you live longer → ~ same number of beats.

HEIGHT IN AUSTRALIA

For the last century, Australians have been getting taller at a bit more than a centimetre per decade. Male and Female Growth Curves are virtually identical to age 11, then the females begin to plateau, slowing their growth to virtually zero around 15–16 years. The males continue their upward trend for about 3 years longer before beginning to plateau.

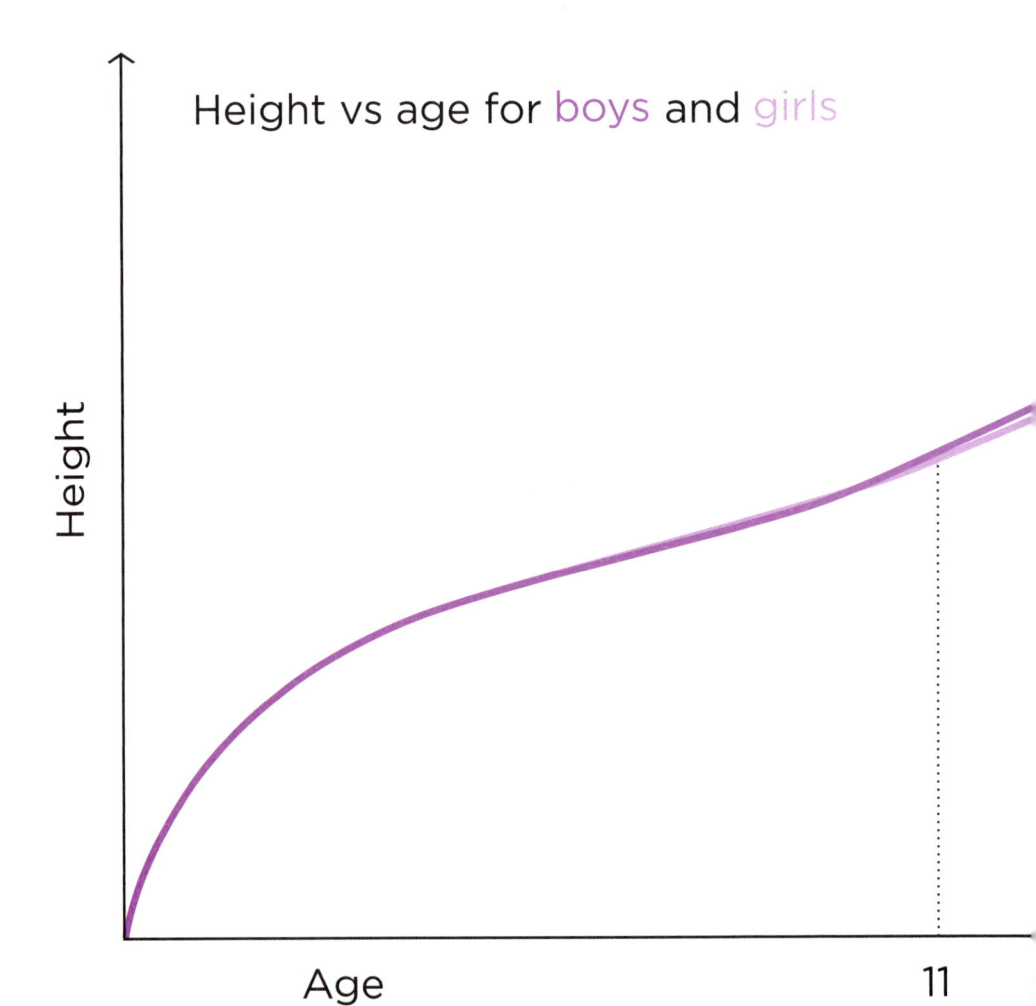

Height vs age for boys and girls

Height

Age   11

A century ago, Australian girls would start their first period about age 16 or 17 – but today, this happens around 12.5 years of age. Australian boys have had a similar drop in puberty age. We're not totally sure why – excess body fat, endocrine disruptors in foods and the environment, less physical activity, stress . . . we simply don't have a clear answer.

Unfortunately, our social systems are falling behind these rapid changes. We are only slowly making changes in furniture and fashion, the timing of primary and high schools, public transport and more.

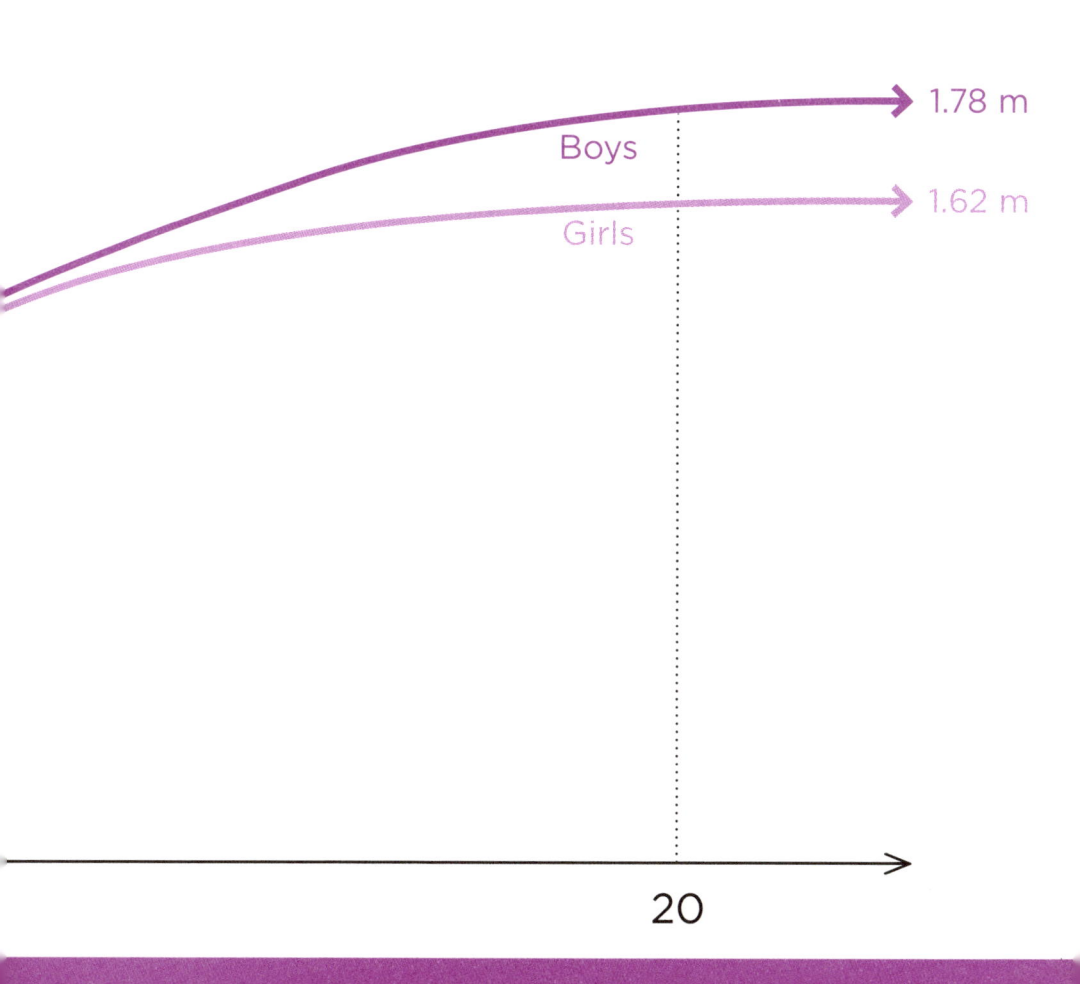

Boys → 1.78 m
Girls → 1.62 m

20

Chilli-eating contest? We each got to having bilateral symmetrical loss of sensation to sides of face, PLUS massive sweating on head.

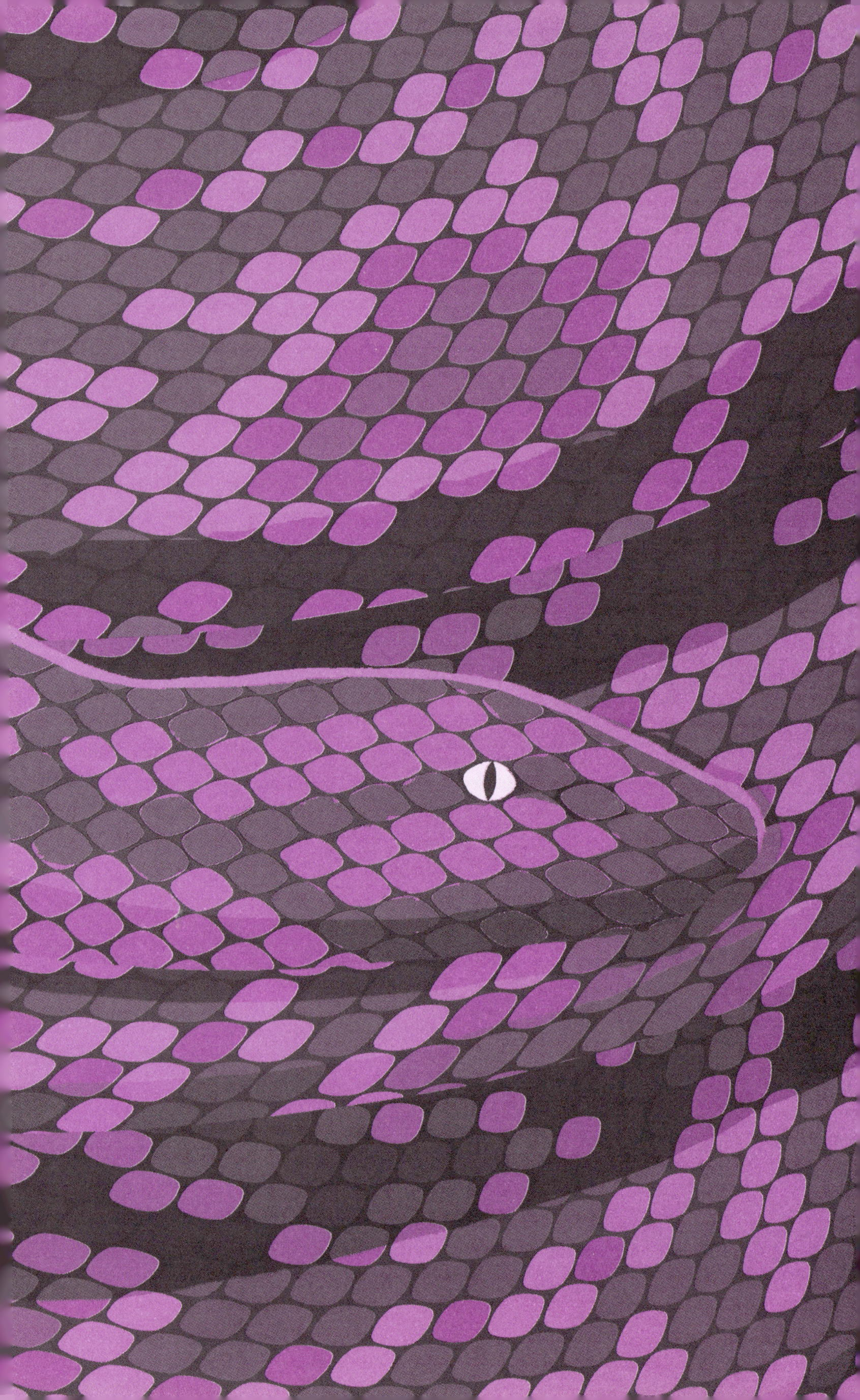

12
PYTHON THE CRUSHER

IN THE ANIMAL WORLD, TOP PREDATORS USUALLY HAVE TO TICK THE RIGHT BOXES.

The checklist starts with eyes on the front of your head to give you stereo vision, so you can judge the distance to your prey. You need good hearing and a sense of smell. It's a given that you have lots of muscles on your trunk and limbs, and weaponry such as big teeth, big claws or big talons.

But what if you have no arms or legs, no ears, no eyelids, and no claws or talons?

Well, in that case, you might fail to be a Top Predator, *or* you might be a python.

Being squeezed to death by a big snake really is the stuff of nightmares. We know a boa constrictor can kill its prey by squeezing it, but until 2015 we didn't know what mechanism actually caused the death.

PYTHON 101

Pythons are non-venomous snakes native to Africa, Asia and Australia. Most of them are "ambush predators". This means that they lie around waiting, then suddenly strike at passing prey, and quickly wrap coils of their own body around it.

Pythons can eat an animal one-and-a-half times heavier than their own weight, work harder than an Olympic sprinter, and starve for up to two years without a feed. How do they survive without food? They burn energy only when it's absolutely essential.

First, unlike mammals, they don't waste energy trying to keep their internal temperature constant. They get warm (and active) from the heat of the day in the daytime, and cool down to lie low at night.

Second, in between feeds, they shrink their gut down to practically nothing. You might not realise it, but it takes a lot of raw materials and energy to run a gut continuously. There are many digestive chemicals and liquids to manufacture and recycle. But once pythons get a meal, they have to suddenly regrow their gut back into existence – it can double in weight overnight. As part of this process, the liver, heart and kidney also increase in mass by about 50 per cent.

In the short term, it takes a huge amount of raw materials – provided by the prey – and hard work to both regrow their gut and digest their prey.

Pythons have to crank up their energy output up to 45 times higher than normal – and keep it there for a few days. You'll appreciate how hard it is when you realise that to win a Gold Medal, an Olympic sprinter will crank up their energy output to 20 times higher than normal – but only for 10 seconds.

@DoctorKarl What happens at night to make lights on the horizon look like they flicker?

So a python just sitting there for a few days, digesting its last big meal, is working about twice as hard as an Olympic sprinter – for about 26,000 times longer, *and* without a break.

THE (OLD) SUFFOCATION THEORY

But how do pythons actually kill their meal?

Well, until 2015, the kind-of-accepted opinion among Herpetologists (Reptile Scientists) was suffocation. (I discuss this in "Python Grows Guts" in my 13th book, *Pigeon Poo, the Universe and Car Paint*.)

It is true that each time the rat (or whatever the python's meal is) breathes out, its ribcage gets a bit smaller. The Old Theory went that at the exact moment when the rib cage shrank, the python would lazily tighten up its grip another notch. After a bunch of tightenings, the animal couldn't expand its ribcage to suck in air – and would quickly die of suffocation. The old (and incorrect) wisdom implied that there was hardly any muscular squeezing – just a lot of taking up the slack and holding on.

But the Suffocation Theory was wrong. The python wasn't just restricting the breathing effort – it was actually squeezing its prey to death.

For example, in many cases, the prey would die quickly – too quickly for the cause of death to be suffocation.

Sometimes there were signs of trauma. A typical trauma case was a poor Malaysian rubber tree tapper. He had been partly swallowed, head first, by a python. Before the snake could suck in the rest of his lifeless body, his corpse was pulled out. The autopsy found multiple fractures in his ribs, neck and the rest of his spinal column.

Maybe pythons could actually squeeze hard...

B/w you & the distant light are many "parcels" of air with differing densities/refractive indices → bend light → flickers.

IF YOU MEASURE, YOU KNOW

This was the background to the research by Professor Scott M. Boback and colleagues. They were the first to elicit the true mechanism of death by python – thanks to modern technology.

In their study, they anaesthetised 15 rats (so they wouldn't feel pain). They then implanted inside each rat various measuring devices to monitor heart rate, blood pressure and the like. They then offered these "robo-rats" to their hungry boa constrictors. On average, the rats weighed about 28 per cent as much as the pythons.

The normal python behaviour was first to strike the rat's head a violent blow with its own head. This usually left the small prey unconscious (but in this case, the rat was already anaesthetised). The snake continued moving forward over the top of the rat, and then quickly twisted to throw two or more loops of its body around the rodent. On average, in this study, the snakes released the lifeless rats after about seven minutes.

The monitors in the rat's body sent the blood pressure and heart-rate data flooding onto the scientists' computer screen. The researchers were astonished to see the boa constrictor could easily squeeze with more than twice the pressure that the rat's little heart could pump. This was curtains for the rat.

@DoctorKarl **How do you think we should approach population growth given limited resources and climate change?**

HEART FAILURE

The heart is a pump. It can't pump "uphill", only "downhill". In other words, like all pumps, it can pump only from higher pressure (generated inside the heart) to lower pressure (inside the rest of the rat's body). In a regular rat, the background blood pressure in the veins is about 4 millimetres Hg, while the heart normally pushes blood out at a pressure of about 82 millimetres Hg. This means that blood can easily leave the heart and travel towards the veins.

But once the boa constrictor squeezed the rat, the internal pressure inside the rat's body increased to 160 millimetres Hg within 6 seconds. There's no way that a pump with an output pressure of 82 millimetres Hg can deliver any blood into a location with a pressure twice as high. Within 6 seconds, the rat's blood pressure had dropped to half, about 40 millimetres Hg.

So much damage was done to the heart that it rapidly went into abnormal electrical rhythms. Even if the boa constrictor had stopped squeezing, the heart would not have been able to pump any more blood.

The various blood measurements (carbon dioxide, oxygen, potassium, pH, etc.) all quickly moved out of the normal range. Indeed, the changes were typical of a heart attack. The blood oxygen levels did drop, but not far enough for suffocation to be the cause of death.

Death quickly followed – which was good for the python. They didn't want to get clawed or bitten by their potential supper.

So the python does crush you to death – not by stopping your breathing, but by stopping your heart. And so, the nightmare reality of death by python is literally heart-stopping...

Population is increasing, but at a decreasing rate. Will reach maximum, then decline naturally. (Education of women.)

EFFICIENT SENSITIVE KILLER

Constrictor snakes are at a disadvantage to venomous snakes.

Venomous snakes can hide in ambush, strike swiftly and inject their venom, and then retreat and wait in safe, concealed comfort until their prey dies.

But constricting your prey means that you have to stay out in the open – and risk attack by other predators. If the prey is still conscious, it could fight back with teeth, claws or talons. Squeezing also burns up lots of energy – seven times more than at rest.

Pythons solve these problems in two ways. Constricting is brutally effective, and they stop squeezing once the prey is dead.

The snake throws its coils around the rat. Within six seconds, the rat's arterial blood pressure has dropped by half. The heart is having great difficulty in supplying enough blood to the rat. Within 60 seconds, the heart rate has dropped by half, and electrical activity of the rat's heart is showing major abnormalities. Very quickly, if the force of the strike hasn't already made the rat unconscious, the abnormal heart activity will do so. The rat is unconscious, so it can't attempt to escape. But it's still alive.

Pythons squeeze their prey for between 10 and 20 minutes. That's a long time to be out in the open – vulnerable to attack. So the python monitors the heartbeat of its prey to know when it's dead.

Another study by Scott M. Boback and colleagues used "warm cadaveric rats" ("cadaver" means "corpse"). They removed the rat's dead heart and replaced it with a "replica heart". This heart was connected to an external pump, so that it could "pulse" and so imitate a heartbeat. Sure enough, if they made the "replica heart" keep beating, the python would keep squeezing. When they switched it off, the python would soon stop squeezing.

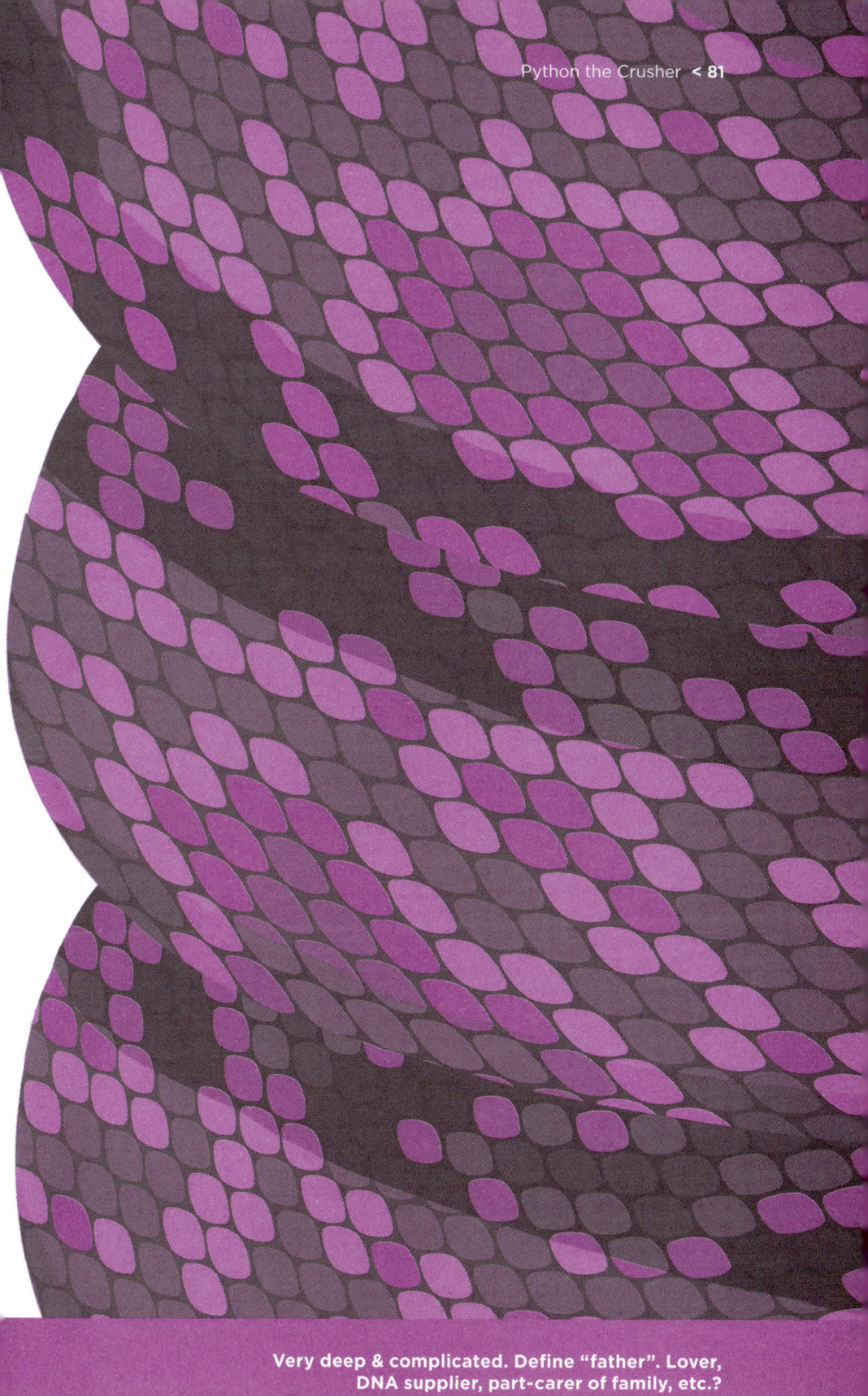

Very deep & complicated. Define "father". Lover, DNA supplier, part-carer of family, etc.?

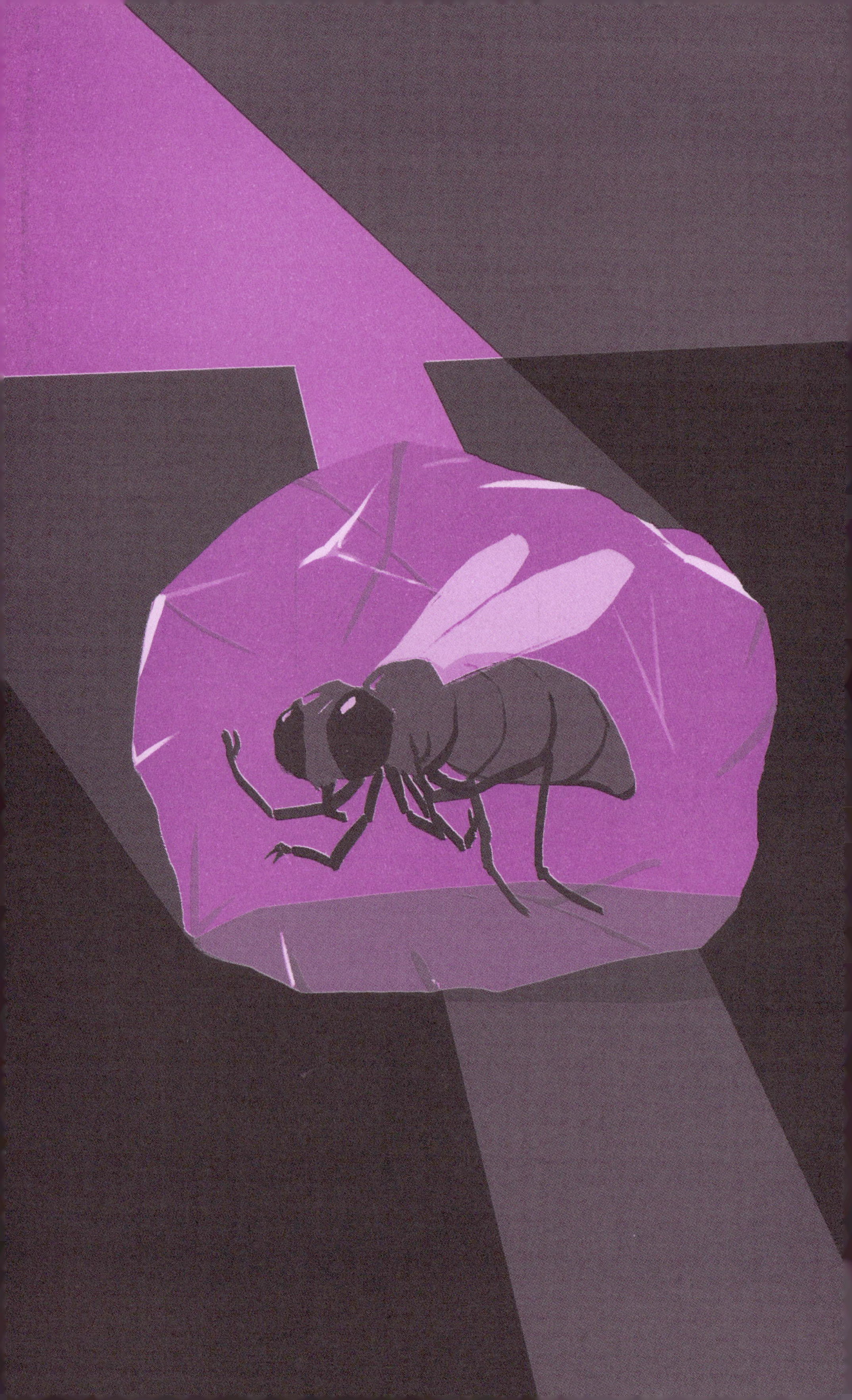

13
FLY EYES AND SOLAR PANELS

ONE OF THE FIRST THINGS THAT I LEARNT AS AN ENGINEER WAS THE OLD SAYING, "NEVER REINVENT THE WHEEL".

Welcome to Biomimetics – where, to avoid reinventing the wheel, we copy Nature to get something we want.

For example, copying the eye of a 45-million-year-old fly can increase the power output of a solar panel by 10 per cent.

BIOMIMETICS

Biomimetics is the rapidly growing science of applying designs from Nature to solve problems in medicine, engineering, materials science and the like.

Professor Andrew Parker, a research fellow at the Natural History Museum in London and a Royal Society Research Fellow at Oxford University, is a leader in this field. Mueller says that understanding "iridescence in butterflies and beetles, and anti-reflective coatings in moth eyes . . . have led to brighter screens for cellular phones and an anti-counterfeiting technique so secret [that Parker] can't say which company is behind it."

Diatoms, one of the most common types of phytoplankton in the oceans, are a surprising source of inspiration. And they're interesting to very different people. Professor Parker has worked with the British Ministry of Defence to copy the waterproof qualities of diatoms and with both Yves Saint Laurent and Procter & Gamble to devise cosmetics that copy diatoms' natural sheen.

Let's leave the ocean for the sky. You usually don't feel the proboscis of a mosquito plunging into your flesh. So perhaps adapting the tiny serrations on the proboscis can make the stab of a vaccination less painful.

Biomimetics can also work at the larger (macro) scale. Perhaps the bumps on the leading edges of the flukes of humpback whales can improve the flight agility of fighter jets? (We can't answer that yet, but they can already increase the efficiency of wind turbines by 15 per cent.) Could the changing patterns of the feathers of raptors in flight help us develop wings that also change shape slightly – to improve both speed and fuel economy?

The thorny devil lizard lives in the Australian Outback. When it hovers its belly over tiny droplets of morning dew, microscopic channels in its prickly hide funnel the water to its mouth within 30 seconds. (I wrote about this in the story "Flat Out Like a Lizard Drinking" in my 11th book, *Absolutely Fabulous Moments in Science*.)

@DoctorKarl Would vegans salivate when they smell lawn mowings?

MULTI-COLOURED CAR

You might have seen a car that looks blue from the front as it comes towards you – but it looks green as it drives away. Yup, it's real, but expensive. How does it work?

Colour comes from many sources.

"Pigments" are probably the one we are most used to. Consider a green pigment. White light (which carries all the colours of the rainbow mixed together) lands on this pigment, and all the colours are absorbed and turned into heat – except for green. Green is reflected, lands in your eye, and you see the colour "green".

Other interesting sources of colour include "interference" and "diffraction". You can have white light passing through very thin films (e.g. a soap bubble) or landing on regular physical microstructures. In each case of "interference", the dimensions are roughly the wavelength of light. Blue light has a wavelength about 0.4 microns (a micron is a millionth of a metre), while at the other end of the rainbow, red clocks in around 0.7 microns.

In the car with the expensive paint, there are tiny fragments of thin-films. As the paint dries, the fragments collect near, and also line up parallel with, the surface of the paint. From one direction, they emit one colour, but from a different direction, a different colour – hence the car paint changes colour as your viewpoint changes.

As another example of a non-pigment colour, the Mayans coated their pots with a paste loaded with microstructures. It produced a bright azure blue colour (Mayan Blue) that hasn't faded over the past thousand years. This is because the colour comes from microstructures that do not change shape over the centuries. This is unlike pigments, which do fade with time.

Nature put diffraction microstructures into living creatures half a billion years ago. They're in oysters, butterfly wings and fossil trilobites.

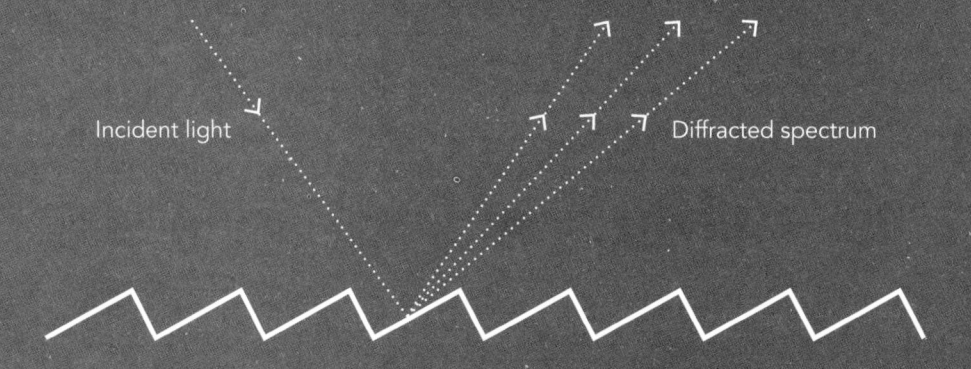

Incident light   Diffracted spectrum

Probably not. Humans can't get nutrition/energy by eating grass (no fermentation chambers). Barbecued eggplant? Yum.

The gecko can run across water, oil, sand and dirt – and still stick to your ceiling. No adhesive tape can do this. But we've learnt from the mechanism that the gecko uses (Van der Waals' forces) and are now making Gecko Gloves that let you climb up the side of a building. (I wrote about this in the story "Atomic Gecko" in my 22nd book, *Bubbles, Bum Breath and Botox*.)

Velcro was invented in 1948, after a Swiss chemist, George de Mestral, became fascinated about how cockleburs clung to his dog's coat. (I wrote about this in the story "Velcro" in my 28th book, *Never Mind the Bullocks*.)

Weight-for-weight, spider web silk is five times stronger than steel – and yet is manufactured at room temperature with water-soluble chemicals. How? We don't know – yet.

Nature has been using evolution for billions of years to make incredible and inventive solutions for plants and animals to flourish.

FLY EYE – ORIGINAL

Some 45 million years ago, there was a type of fly that was active in the dim light around dawn and dusk. One of these unfortunate flies got caught in the slow-flowing sap of a tree, and ended up being almost perfectly preserved in what became a block of solid amber.

Some 45 million years after it died, modern scientists looked at it with a high-powered Electron Microscope.

They noticed some very fine regular parallel corrugations on the front of the fly's eyes. These corrugations were a regular 250 nanometres apart – about half the wavelength of blue light.

It turns out that often, very fine corrugations or lines can, via the Physics phenomena of diffraction or interference, produce beautiful colours. For example, there are similar very fine structures on the wing of the *Morpho* butterfly, which lives in Central and South America. These structures create flashes of blue light that have been seen from aeroplanes flying overhead.

But the fly eye wasn't flashing anything. No colours were being emitted – in fact, the opposite – hardly any light at all was being reflected. The anti-reflective action of the fly eye corrugations covered a wide range – over 120° on the front of the fly's eye. Virtually all the light landing on the front surface of the fly's eye was entering the eye. This is the first known anti-reflective optical device, 45 million years old.

Now think of the last solar panel you looked at. Did you catch a shiny reflection off the glass at the front, as you looked at it from different angles? Probably yes. So here we can learn from Nature – again. The light from the solar panel that landed inside your eyeball was being wasted. It should not have reflected off the front of the solar panel. No, it should have gone into the solar panel and been turned into electrical power.

FLY EYE – COPY

And that's exactly what our Biomimetics engineers have done. The principle was simple, but making it work was tricky.

First, they worked out the refractive index (or "bending power") of the glass on the front of the solar panel.

Then they had to find a glue that was both totally transparent and had exactly the same refractive index as the glass. Furthermore, it had to be able to survive for decades under direct sunlight.

And then they had to find a plastic with the same properties (totally transparent, same refractive index as the glass) that would also survive for decades. There was an extra engineering requirement. It had to be soft enough so it would be relatively easy to machine with the very fine regular corrugations, and yet hard enough so that the corrugations didn't degrade over the decades.

The engineers succeeded in gluing this plastic with fine corrugations onto the front of the glass of the solar panel.

And so now, thanks to copying Nature, we have solar panels that absorb much more of the light that lands on them, reflecting none. So they produce 10 per cent more power.

Lottery? 1) Lottery is a tax on mathematically innumerate 2) Your chances are only slightly improved if you buy a ticket.

NATURE'S METHODS

But we want to do more than just ape the products that Nature makes – we want to copy her manufacturing methods.

Think about the plastic glued onto the front of the solar panel. It takes a lot of temperature and pressure, and fancy manufacturing processes, to end up with this finely corrugated plastic. Yet Nature makes the corrugated fly eye surface with organic chemicals at room temperature and pressure – very impressive.

We know that catalysts and some kind of fancy self-assembly have to be involved – but we don't know the fine details, yet.

Self-assembly? Think of how nature assembles the tiny electric motors built into the cell walls of bacteria (see "Electric Motors in Bacteria" on page 245). That process seems unlikely, but Nature does things like this all the time.

And when we try to make nanostructures, all we can manufacture are tiny fragments a few millionths of a metre across – but Nature makes nanostructures centimetres across.

One possible pathway is to get biological cells to grow these structures for us. After all, biological cells made the original 45-million-year-old fly's eye.

But where do we start? There are so many natural structures to copy.

Let's go for something very useful – a structure that can catch water from the air.

@DoctorKarl How do veins/arteries know how much blood the muscles need to survive?

BEETLE TO DRINKING WATER

Delivering water is a very worthwhile project. Ten per cent of humans don't get clean drinking water.

It turns out that the Namib Desert in Africa, one of the hottest and driest environments in the whole world, supports the tenebrionid beetle. It carries beautifully evolved microstructures that harvest its drinking water – from the very air around it.

The beetle faces into the morning fog coming off the ocean and lifts up its back end to an angle of about 45°. There are little bumps on the beetle's back end that are hydrophilic (that is, water-loving). They capture tiny droplets of water, which coalesce into bigger drops. Once the drops get big enough, they roll downhill into special channels that are hydrophobic (that is, water-hating). These channels don't allow the water to stick. Gravity continues to influence the water, and the drops roll downhill all the way to the beetle's mouth. Professor Andrew Parker reckons he can build structures that can collect one litre of water per square metre per hour – by copying these structures from Nature.

With fly eyes and beetles' backs, insects might not take over the world, but instead help save it by showing us how to make clean power and clean water.

Over the millions of years, Evolution has matched supply & demand of blood vs end organ usage.

14 CEMENT SHOES

So-called "cement shoes" have long been an iconic image from the murky world of the Mafia. It's a terrifying concept – your feet are forced into a tub of cement, and then you're dumped into a watery grave, to vanish forever.

But there had never been a rock-solid case proven, until 2 May 2016. A uni student walking near Manhattan Beach in Brooklyn, New York, came across the corpse of Peter Martinez. The feet were buried up to the shins in a five-gallon (19 litre or about 45 kg) bucket filled with concrete – a mixture of cement, gravel, sand and water. Peter Martinez, known as "Petey Crack", had over 30 convictions. (He was aka "Bad Petey".)

There had been rumours about the untimely deaths of various criminals over the years. One of the earliest stories related to the disappearance of Danny Walsh, a bootlegger from Rhode Island. According to an Associated Press clipping from 3 June 1935, he "was stood in a tub of cement until it hardened about his feet, and then thrown alive into the sea". But this was just hearsay – the body was never found.

There were proven cases of blocks of concrete used to weigh down corpses. This happened to both Johnnie "Chink" Goodman (Philadelphia racketeer, found in a creek in New Jersey, 1941) and Ernest "The Hawk" Rupolo (hit man, thrown into Jamaica Bay in New York, 1964).

But I have a few quibbles with the cement shoes story – and I'm not just objecting to the violence.

First, "cement" – as in "cement shoes", "cement overcoat" and "cement boots" – is wrong. It should be "concrete". Cement is the binder that sets and hardens, joining the gravel, sand and water. Concrete is the combination of the cement with these components to make the world's most popular building material. There's more than a cubic metre made per person per year, most of which is not used by the Mafia (as far as we know). (OK, I know I was being pedantic.)

Second, it can take up to 12 hours for the concrete to harden so much that one cannot remove one's feet. So either the feet have to be inserted after the victim is dead or somebody has to watch the live victim to make sure they don't remove their feet.

And how did Peter Martinez's body wash up, with many kilograms of concrete weighing it down? Well, the waves were very strong that day, but more importantly, the concrete had been poorly prepared. It was full of air bubbles. As they say, "You can't get good help nowadays."

Once again, mobile phones don't cause brain cancer: theconversation.com/new-study-no-increase-in-brain-cancer-across-29-years-of-mobile-use-in-australia-58927 Nor do they cause hand cancer (you hold phone with hand).

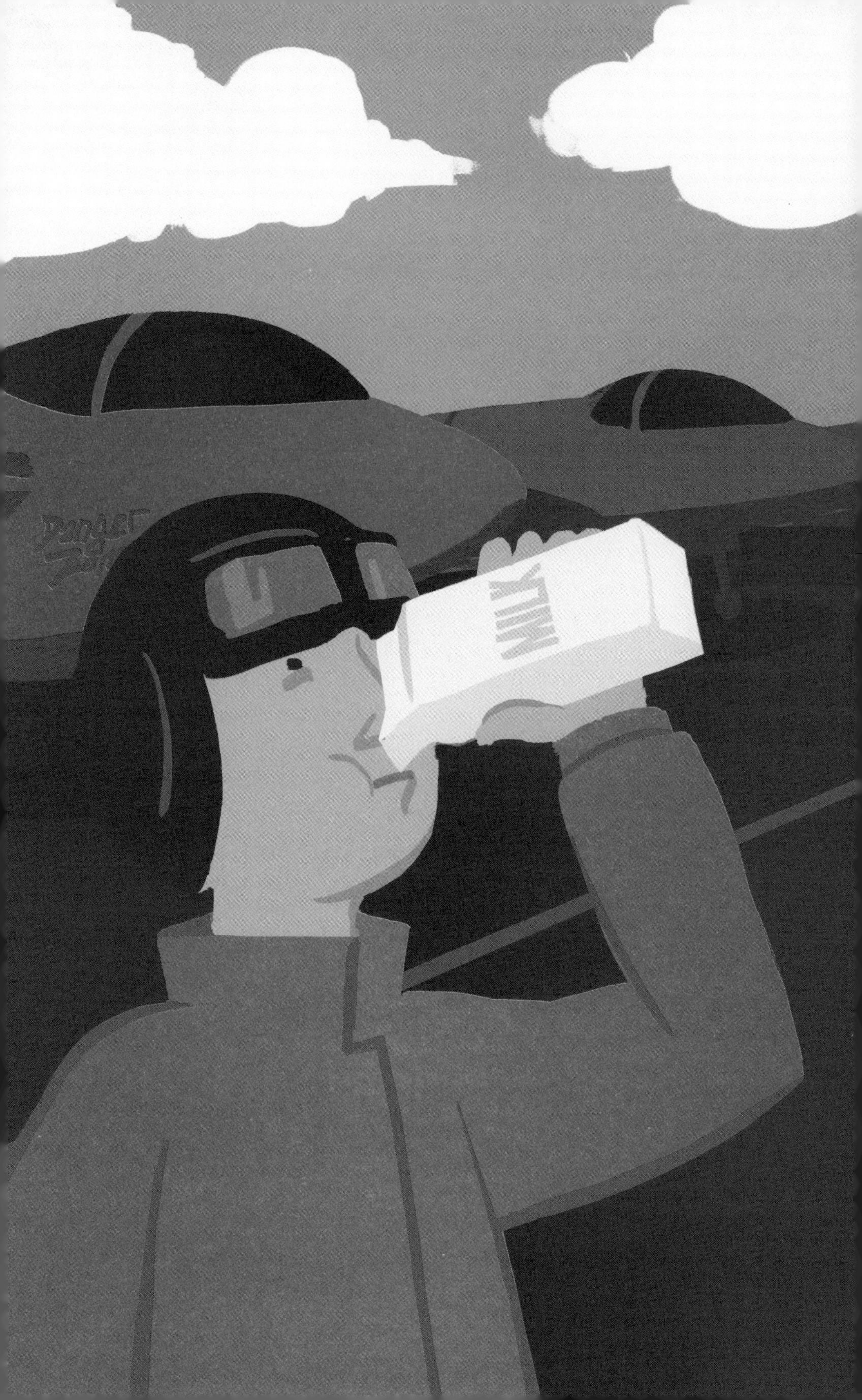
MILK

15

HOW MANY CELLS IN YOUR BODY?

THE QUESTION "HOW MANY CELLS ARE IN THE HUMAN BODY?" IS DEEPLY FUNDAMENTAL TO OUR BIOLOGY. IT SEEMS AS THOUGH WE SHOULD HAVE WORKED OUT THE ANSWER DECADES OR CENTURIES AGO.

But surprisingly, it's only in the past few years that we've even approached a reasonable estimate.

Current best answer? About 37 trillion – that's 37,000 billion, or 37 million million, or 37,000,000,000,000 or 3.7×10^{13}. And they're mostly Red Blood Cells.

(Just to get a handle on Big Numbers, the Gross Domestic Product (GDP) of the entire planet is US$70 trillion or so.)

CELLS – ONE TO MANY

Bacteria have just one cell. That's it, just one cell for eating, drinking, moving, excreting, making chemicals – everything.

A tiny transparent worm (*Caenorhabditis elegans*) is somewhat famous for being the first multicellular animal to have its total genome sequenced – we know all of its DNA.

A REASONABLE ESTIMATE

Let me give you an idea of what I mean by "reasonable". Going back over the last two centuries, the estimates for the number of cells in the human body have been in the trillions. How many trillions? Anywhere between five trillion and 200 million trillion. That's a range of 40 million to one!

An estimate that has a range of 40 million to one is pretty close to useless. Imagine going to the shops to pick up some milk, and being unsure as to whether to get one litre (about one kilogram) or 40 million litres. Forty million litres weighs about 40,000 tonnes – the mass of a medium aircraft carrier, about a quarter of a kilometre long.

Let me give you an idea of what scientists mean by "reasonable". Of course, this will depend on what Field of Science you consider – Psychology, Physiology, Physics, etc. But as a rough rule of thumb, when you're doing new research, any result that is in the "right Order of Magnitude" is reasonable. ("Magnitude" in this context means a "power of 10".) Suppose that your rough calculations tell you to expect a value of 50. Well, in the early days of your research, any result bigger than five and smaller than 500 could be seen as "reasonable". Of course, if both your Theory and Experimental Methods are okay, you should pretty quickly get close to 50.

@DoctorKarl How come fish survive lightning strikes in water?

The adult male *C. elegans* has exactly 1031 cells. The adult hermaphrodite is made from just 959 cells. (There are no female *C. elegans*.)

And humans have about 37 trillion cells.

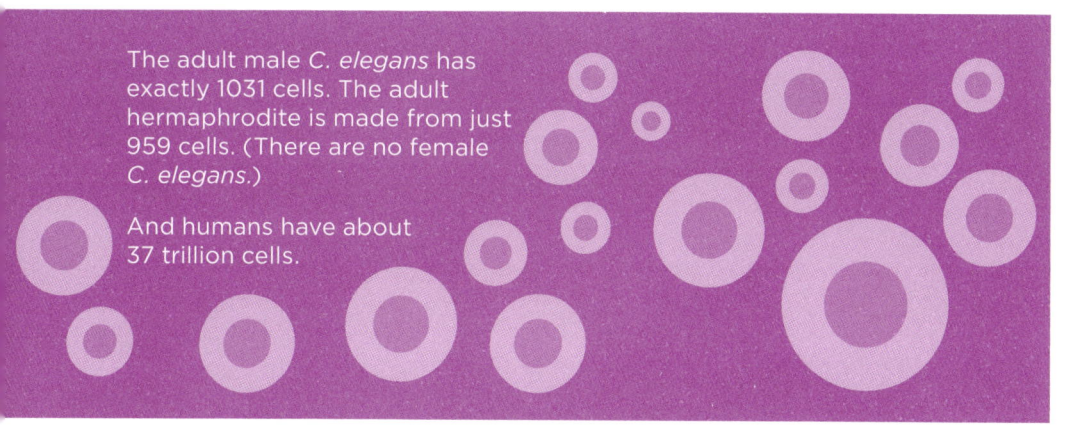

CELL 101

If you haven't studied any Life Sciences (like me, back when I was a Physicist and Mathematician), you might not be familiar with "cells".

Your body has about 200 different types of cells. An "average" cell weighs about one nanogram, and has a volume of about four billionths of a cubic centimetre.

We first saw cells in 1665, about three-and-a-half centuries ago, when the British scientist Robert Hooke peered into an early compound microscope. He sketched the amazing detail he saw in some 60 objects he chose at random.

One object he examined was a very thin slice of cork – the rubbery wood that is still sometimes used to seal bottles of wine. He saw thousands of tiny pores that he called "cells", from the Latin word "cella", which meant a small room, like the ones that monks lived in.

Robert Hooke didn't realise at the time that these empty "cells" had previously been filled with living tissue.

It took nearly two centuries before Theodor Schwann and Matthias Schleiden formulated what we now call "Cell Theory" – around 1838. Today, modern Cell Theory is part of the bedrock of Biology.

They do die, but only in a small (~5m) circle. Not a lot of data. Based on anecdotes from Tropics, e.g. Malaysia.

RED BLOOD CELLS

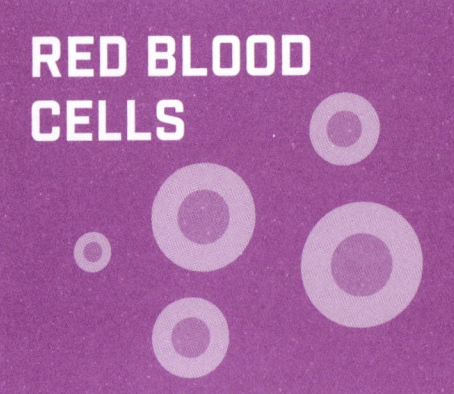

Red Blood Cells make up about 45 per cent of the five litres of blood that circulates in the average adult male. And yet this small mass (less than 2.5 kilograms, or about 3 per cent of total body weight) makes up about 26.3 trillion cells – about 70 per cent of all the cells in the body.

How can this be? The answer is that Red Blood Cells are very small.

Cell Theory includes the following seven statements.

First, all living creatures are made of one or more cells. That covers the enormous range from bacteria to the blue whale, the biggest creature we know of.

Second, the cell is the fundamental unit of structure and function in all living creatures.

Third, all living cells come from previous cells, which grew and then divided.

Fourth, the overall activity of a creature depends on the total activity of all the individual cells.

Fifth, energy flows into cells and inside those cells, and can also flow out of cells. Think of them as little batteries.

Sixth, cells contain hereditary or genetic information called DNA. This DNA is passed from one cell to its daughter cells. (By the way, in one survey, 8 per cent of Americans said they don't accept the existence of DNA. In a different survey, 80 per cent wanted food that contains DNA to be labelled as such. They don't seem to realise that both meat and vegetables contain DNA).

And finally, all cells in similar species have the same basic chemical composition.

They have no DNA because they are fiercely optimised for their task of carrying oxygen.

And this leads to another surprising result.

Only 30 per cent, or 10.9 trillion of your human cells, carry your DNA. The remaining 70 per cent (Red Blood Cells) don't.

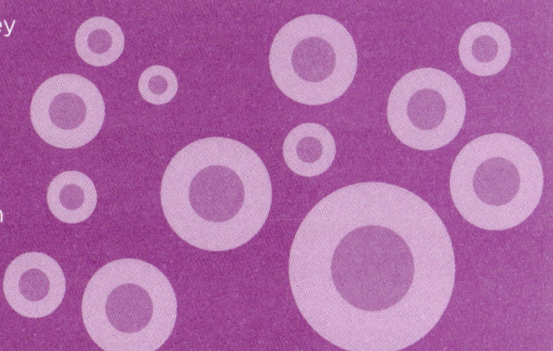

CELLS IN YOUR BODY?

That's all very good, but how many cells are in your body?

Back in 2013, Dr Eva Bianconi from the University of Bologna in Italy, and colleagues from Italy, Spain and Greece, tried to answer this question.

The team first searched for estimates in the medical and scientific literature. These estimates came from fields such as Biology, Histology (or Micro-anatomy), Anatomy and Physiology over the last two centuries. The estimates were all in the trillions. There was one very extreme estimate of a colossal 200 million trillion cells in the human body, but they excluded this outlier.

CELL NUMBERS CHANGE?

The numbers of cells do change in a few known ways.

For example, a healthy liver would have about 240 billion cells, while one suffering from cirrhosis would have about 28 per cent fewer, at 172 billion.

The number of Red Blood Cells will vary a little from males to females, as well as in pregnancy, with age, and in populations living at high altitudes.

Each hour of your life is a progressively smaller percentage of your total number of hours lived → time SEEMS faster.

Most of the estimates lay between five trillion and 70,000 trillion. That's still an enormous range: 14,000 to one. That's the difference between one litre of milk, and two thirds of the load carried inside those 6-metre containers you see on the back of semi-trailers.

The team went back to First Principles – back to basics.

First, they defined the average human to be a 30-year-old male, 1.72 metres tall, weighing 70 kilograms, and with a surface area of 1.85 square metres.

Second, they got their estimate the hard way. They looked at some 19 individual organs or systems in the body, and worked out how many cells were present in each organ or system.

But while Red Blood Cells can be relatively easy to count, nerve cells often weave themselves into a tangled web. Another problem is that our cells are not packed evenly, like a bottle full of spherical marbles. Human cells come in vastly different sizes, and can be packed tightly or loosely.

Even so, the researchers persevered, and finally just added up the numbers.

Of course, it was complicated.

19 ORGANS OR SYSTEMS

The team painstakingly worked their way through the alphabet, from A to V. Adipose (fat) tissue was followed by articular cartilage, the bilary system (gall bladder), blood, bone and so on.

For example, for bone, it's fairly easy to start with the total weight of the skeleton. But there are two different types of bone. Trabecular tissue (23 per cent of bone tissue) is the softer, spongy inner bone which contains the bone marrow, while cortical tissue (77 per cent) is the harder, outer compact layer. So the team came up with 700 million cells of trabecular bone tissue, and 1.1 billion cells of cortical bone tissue.

Eventually the researchers ended up at V for vessels (such as veins and arteries).

The blood vessels include about 80,000 kilometres of capillaries,

@DoctorKarl **Does a room with the lights on get darker if the door is opened, into a black space?**

with an average diameter of three quarters of a millimetre. Blood vessels are not pipes made from inorganic dead material. Blood vessels are more like a whole bunch of flattish cells that first stick together to make a rectangular sheet, and then roll themselves into a tube. On the inner hollow side of the tube, blood vessels are lined by endothelial cells. Individually, these endothelial cells are only about 60 microns (a micron is a millionth of a metre) long, and 20 microns wide. That gives you about 2.5 trillion endothelial cells in the capillaries alone.

It turns out that just six cell types account for 97 per cent of our human cells. The Red Blood Cells make up 70 per cent, glial cells (support cells for the neurons) 8 per cent, endothelial cells (lining blood vessels and airways) 7 per cent, dermal fibroblasts (a kind of skin cell) 5 per cent, platelets (cells that help blood clot) 4 per cent and bone marrow cells 2 per cent. There are not many muscle cells (240 million, or 0.0006 per cent), but they are very large, so they make up about 20 per cent of your weight.

Adding all these numbers together gives a total of 37 trillion cells. Mind you, with further work, this will almost certainly be adjusted a bit – but hopefully, not by a factor of 14,000! So we can count on that for a while.

Hopefully, Dr Bianconi and her team didn't need a stay in a padded cell, after all that endless cell counting...

MORE CELLS OR GALAXIES?

Are there more cells in a human, or galaxies in the heavens?

Well, how do you work out the total number of galaxies? The simple first approximation is to point a really sharp telescope unblinkingly at the same bit of sky for a week or so. Count the number of galaxies. Multiply to get a reasonable approximation to the total number of galaxies. The result is between 100 and 200 billion galaxies.

So there are about 200 times more cells in your body than there are galaxies in the visible Universe. (But then, each galaxy has about 200 billion stars.)

Yes. A small percentage of the fixed number of photons/sec from light are no longer reflected back into room, when door opens.

NON-HUMAN CELLS: MY BAD

Yep, once again, I am clearing up a mistake. In 2011, in my 31st book, *Brain Food*, I (wrongly) wrote that we are host to about 10 times more bacterial cells than human cells in our bodies. Incorrect – there are roughly equal numbers of human and bacterial cells.

Dr Ron Sender and colleagues worked out how this incorrect info made its way into the common knowledge – and yes (shockingly) even into the peer-reviewed literature.

Way back in 1972, Dr T. Luckey wrote a paper, "Introduction to Intestinal Microecology", in the *American Journal of Clinical Nutrition*. Dr Luckey was one of the first people in the field who tried to work out the overall number of bacteria in the human gut. So he did a rough Order-of-Magnitude, back-of-the-envelope estimate by first assuming that there were about 100 billion bacteria in each gram of liquid in the gut. (There was some support for this.) He also assumed there was about 1000 grams of liquid in the gut. That gave about 100 trillion bacteria in the gut.

This was then quoted by Dr D. Savage in the *Annual Review of Microbiology* in 1977, and then re-quoted in other journals.

In 2005, F. Bäckhed et al. in *Science* quoted the same number of gut bacteria, but reduced the human cell count: "the population – up to 100 trillion – is about 10 times greater than the total number of our . . . germ cells".

Now the incorrect data (that we carry 10 times as many bacterial cells as human cells) had wormed its way into the literature. It stayed there and was continually re-quoted.

And that's how I made my mistake.

But here's a thought. Each time you have a large bowel motion, you might expel enough invading bacterial cells to regain numerical supremacy for your human cells.

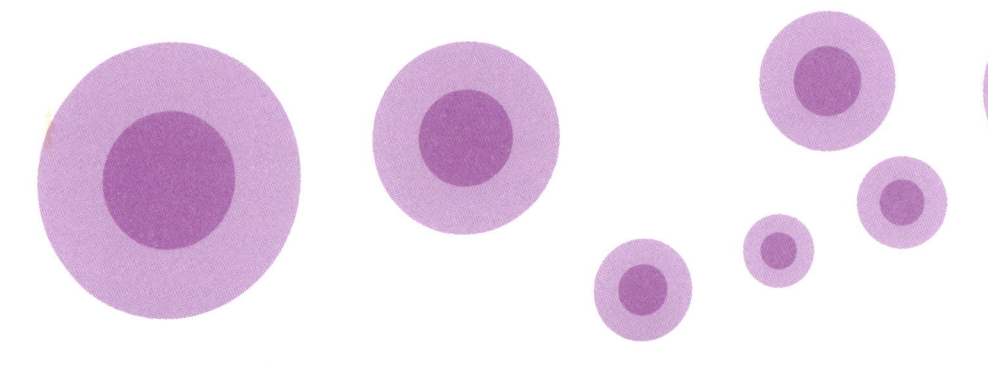

@DoctorKarl Would adrenaline rush from jumping out of plane be able to overcome an allergic reaction?

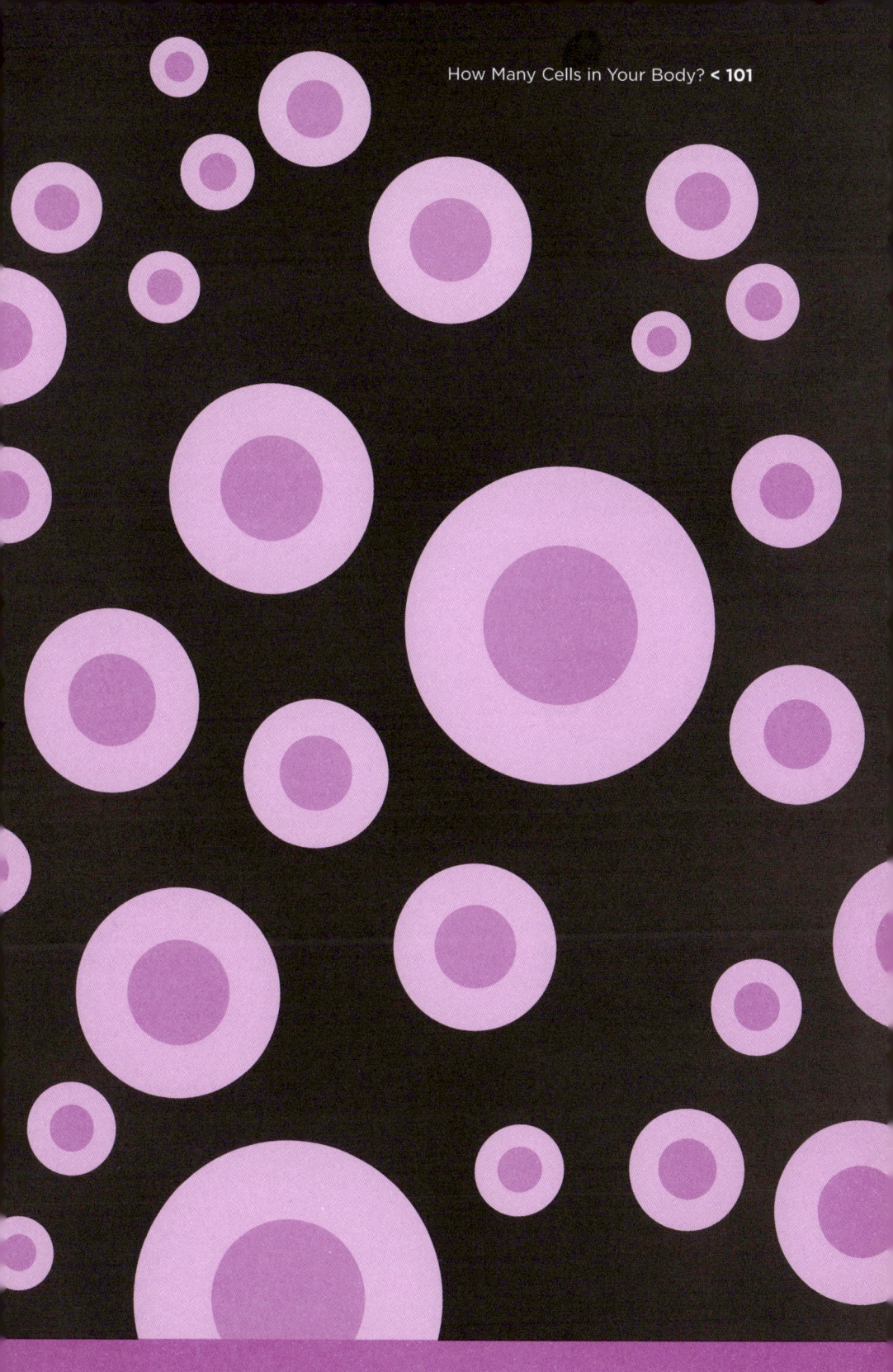

1) Almost certainly not 2) Difficult to get this study past an ethics committee 3) With or without parachute?

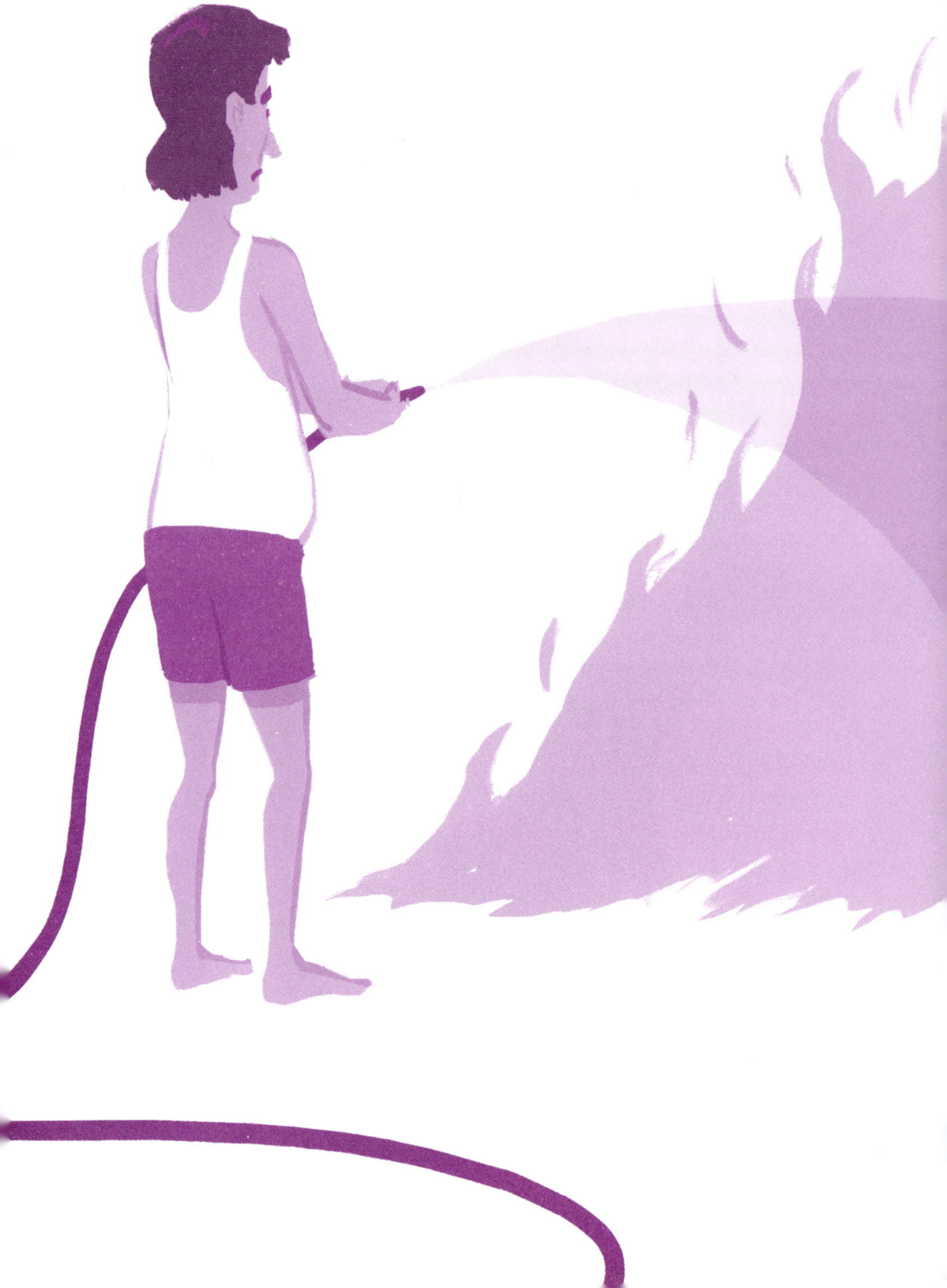

16

WATER BURNS PLANTS?

I'VE HEARD MANY TIMES NEVER TO WATER THE GARDEN PLANTS UNDER THE POWERFUL GLARE OF THE MIDDAY SUN. IT'S A LONG-ACCEPTED TRADITION COMMON TO BOTH AMATEUR AND PROFESSIONAL HORTICULTURALISTS.

The basic gist is that a water droplet sitting on a leaf acts like a tiny magnifying lens, and can focus the sunlight to scorch and burn the innocent leaf.

HOW?

Every now and then, I read an article in a newspaper that reports on a Scientific Finding – and which is Wrong.

So I wasn't surprised when I read a Press Release published under the banner of the European Commission. It reported on, and quoted, research that showed that watering plants at midday overwhelmingly did not

On the interwebs, about 80 per cent of sites that discuss this claim reckon that water plus Sun can equal leaf burn.

Finally, some scientists have run experiments to check out the theory of watering in the midday Sun – and here are the results.

Water droplets on smooth leaves don't cause burns from the Sun, but it might just be barely possible to burn a hairy leaf.

GLASS BALLS BURN

Dr Adam Egri and colleagues from Hungary and Germany carried out both computer modelling and tests on real leaves. In their experiments, they used water droplets and, as a comparison, little glass balls.

In their first study, they covered the surface of a smooth leaf with glass balls ranging in size from 2 to 10 millimetres. They exposed them to the Sun for periods between one and nine hours. And sure enough, there were scorch marks.

So it would seem as though there was some truth to the claim that sunlight passing through water droplets could burn leaves.

But while glass and water are each transparent, they have significant differences. For one thing, glass balls keep their shape, unlike water droplets that spread out and flatten. The other difference is that glass bends light about 15 per cent better than water does.

trigger sunburn to their leaves. Inexplicably, the title of the Press Release said the exact opposite: "Watering Plants Midday Triggers Sunburn, Research Shows". How ridiculous . . .

WATER ON SMOOTH LEAVES – MIDDAY

So the second study examined water droplets on smooth leaves. But, no matter how the scientists altered various factors, they could not get the sunlight shining through the water droplets to burn the smooth leaves.

When the sun was shining from directly above, the light would be brought to a focus underneath the water droplet, inside the body of the leaf. But there was a big blob of water sitting on top of the leaf, soaking up the heat. Any heat that got into the leaf was immediately transferred to the water. (Water has a large Thermal Mass – it can soak up lots of heat energy.) But, as the water got warmer, it evaporated itself out of existence. And once there was no water droplet, there was no focusing of the heat. So there were no burn marks on the leaf in the midday Sun.

(For completeness, they should repeat the study in a very humid environment. The humidity could delay evaporation, and the droplet might get quite warm, or hot.)

More contrails in USA? Related to 15x greater population, and more air travel?

WATER ON SMOOTH LEAVES – NOT MIDDAY

What about sunlight in the morning and afternoon, when the slanting light is beaming through the water droplet from the side, and landing on the bare leaf just outside the water droplet? Could that cause a burn?

The answer turned out to be "no".

Let's start by considering the shape of the water droplet.

It turns out that different leaves have different degrees of "hydrophilia" or "water-loving" tendencies. So the scientists tested leaves from various trees – maple (very water-loving), plane (less water-loving) and rowan (water-hating, or "hydrophobic").

At the water-loving end (maple), the water droplets were very flattened, while at the water-hating end (rowan), the droplets were fairly spherical. As you would expect, the more spherical water droplets on the water-hating leaves did concentrate the heat better than the less spherical droplets. There was indeed a slight "hot spot" of light just outside the droplet, landing on the leaf. But it was not hot enough to scorch the leaf.

One reason that the leaf didn't get too hot is that just under the surface, leaves have a very efficient system of horizontal channels to carry unwanted heat away. (Plants have had nearly half a billion years to evolve clever ways to deal with temperature extremes.)

Another reason is that at the start and the end of the day, there is more air for the sunlight to travel through – which reduces the intensity of the direct beam. When the Sun is just above the horizon, the sunlight is travelling through about 22 times more air than when the Sun is directly above you, at the vertex, around midday. This is called "Atmospheric Extinction". For the same reason, at night, a faint star that is just visible when directly above you will often fade from view as it drops to the horizon.

Getting back to the morning and afternoon Sun (at 23° elevation): there's not enough heat to burn. So sunlight shining through water droplets doesn't burn smooth leaves.

@DoctorKarl **What determines the brightness of Auroras?**

Water-loving maple tree leaf – flattened droplet

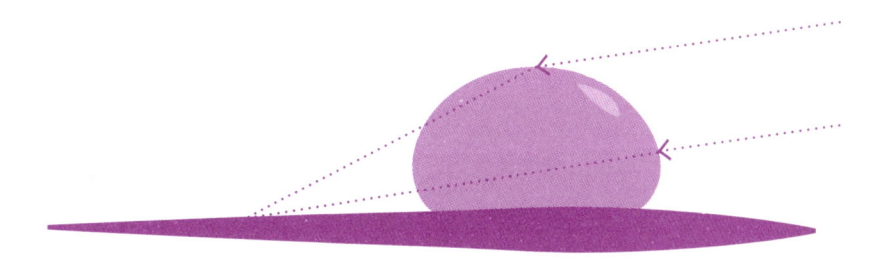

Water-hating Rowan tree leaf – spherical droplet

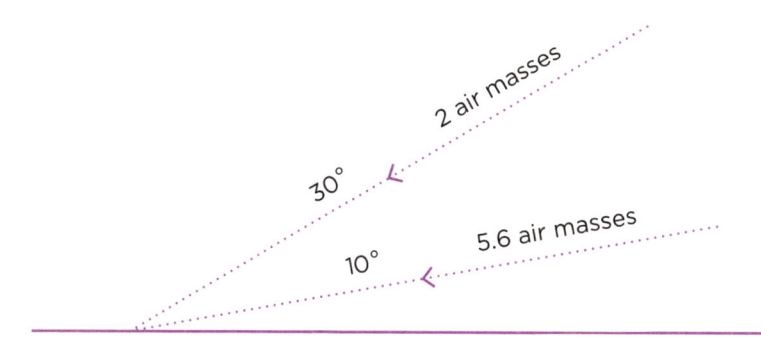

30° 2 air masses
10° 5.6 air masses

WATER ON HAIRY LEAVES

But what about the hairy plants, such as lotus or floating fern?

Their leaves have hairs that are "hydrophobic" or "water-hating". Under the right circumstances, these hairs could catch a water droplet and keep it suspended at the right distance above the leaf to allow the focus point of the sunlight to land on its surface. Because the water is not in contact with the leaf, there is no cooling effect. And the hairs,

Aurora = God's TV set (e.g. old CRT TV). High-energy charged particles from Sun hit O_2 (green often) & N_2 (red often).

being water-hating, would repel the water, and so keep it spherical – the best shape to focus the sunlight.

Still, the burn usually does not happen.

The hairs hate water, so the merest breath of wind shakes the droplets off. Furthermore, the smaller droplets are better at focusing the heat – but being smaller, they would also evaporate very quickly.

But, the scientists did sometimes see scorch marks produced by water droplets on hairy leaves.

EVOLUTION WINS AGAIN

How did this Don't Water Plants At Midday story arise? Who knows? (Maybe it was tied into the old saying that "only mad dogs and Englishmen go out in the midday Sun".)

We do know that water can hold chemicals (both natural and artificial). When this water evaporates off a leaf, it can leave a residue – one that superficially resembles a burn. But it's not a burn, because you can easily wipe off the residue with a wetted finger.

Perhaps the real reason that you should avoid watering in the heat of the day is that the water will more readily evaporate away. This leaves less water to go into the soil and the plants. In dry countries, this is a really big issue.

But how's this for a simple debunking argument? Plants have adapted to every conceivable ecological niche. You'd think that after half a billion years of evolution, they would have learnt to deal with a little rain, even if it falls at midday . . .

WATER BURNS SKIN?

Many sites on the interwebs make the incorrect claim that water droplets at midday can burn plants. So are there similar claims for human skin?

According to Dr Egri and colleagues, yes.

They looked at nine dermatological and cosmetics websites (mostly non-English) that considered the possibility of "sunburn of the human skin due to sunlight focused by water drops during sun-bathing. Can sunlit water drops burn the human skin? The rate of the 'yes' answer was $8/9$ = 89 per cent."

The interwebs are not the cause of the wrong information. They just let the misinformation travel across the Earth much more quickly than when books and newspapers were the predominant methods of spreading information.

Yes indeedy. A sign of "weak" coffee addiction.

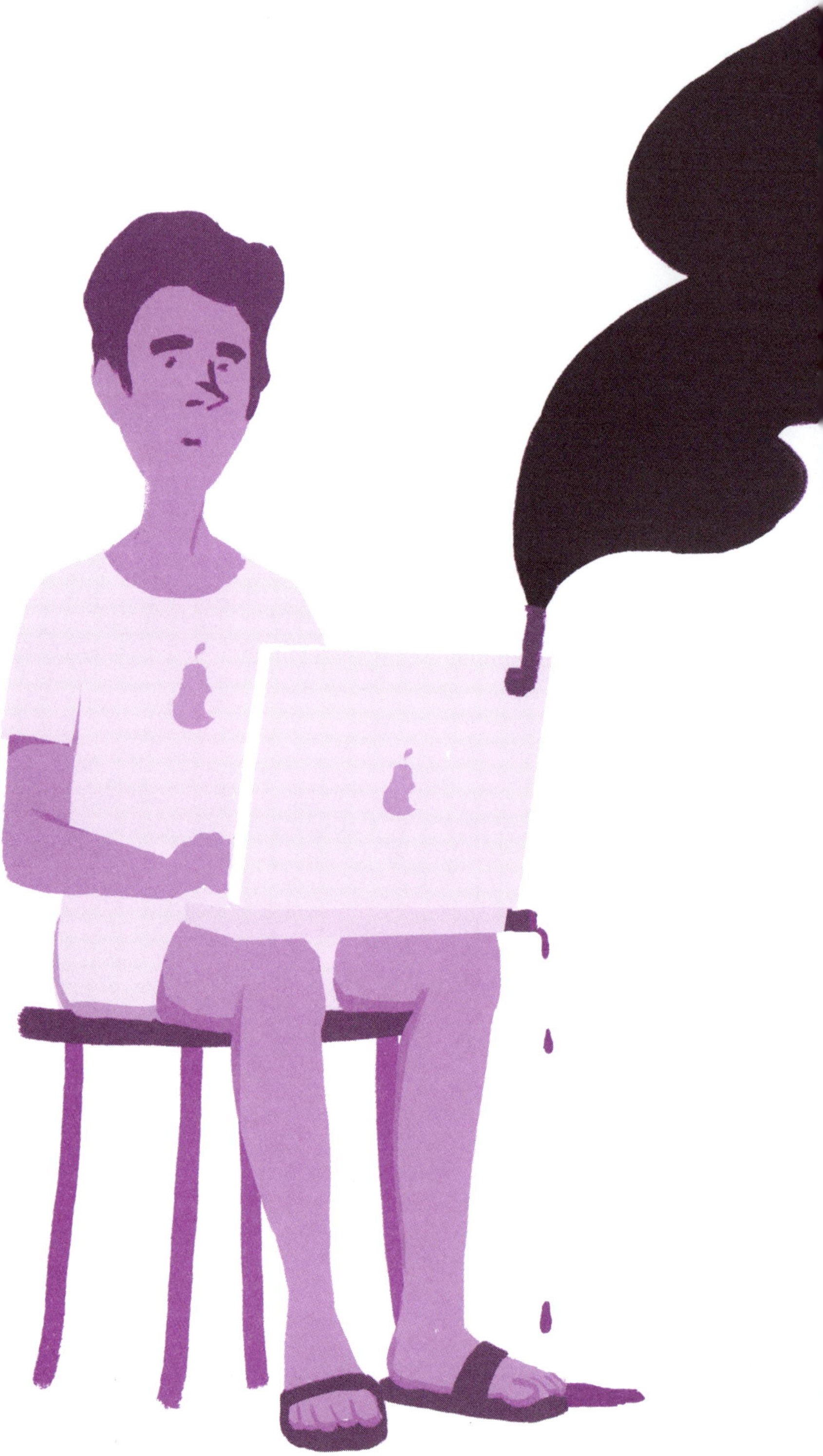

17

DIRTY DATA

IT SEEMS PERFECTLY OBVIOUS THAT GOING DIGITAL MEANS WE MUST BE CHANGING TO A CLEANER AND GREENER ECONOMY.

Your new smartphone, or tablet, or computer, looks so shiny and clean. And surely telecommuting from home must be cleaner in terms of Big Picture Energy Consumption. You're not shifting lots of heavy atoms from here to there, you're just shifting "weightless" electronic bits. After all, you're not burning energy by travelling to an office – you're just sitting at home.

Is the new economy cleaner and greener? Yes and no. It all depends on how we manufacture, and maintain, our new internet-enabled world. Unfortunately, so far, we've had a pretty dirty start.

CLOUD 101

The days are long gone when computers were standalone devices – today, they're "connected" via the net. In the near future, the internet will probably be the biggest thing we humans will build. Following current trends, by 2020, 80 per cent of all people on earth will be connected to the net, and there will be some 7.6 billion mobile broadband subscriptions.

Fundamental to the expansion of the internet is the so-called "Cloud". This mysterious Cloud is remote data or services that we access via the net. It's a truly "disruptive" technology. Increasingly, the Cloud is where we store our photos, manage our health, do our work, embark on human relationships, go shopping online, book flights and accommodation, send email and watch movies. Each time you use your smartphone, you are accessing some hundred or so other computers around the world.

The Cloud is not a diffuse, vaporous entity. "Data centres" – actual physical buildings – store our invisible Cloud data. These data centres are built in areas that are stable – free from civil unrest, earthquakes, floods and the like.

But more importantly, data centres need high-speed optic fibre communication (which is relatively cheap), and electricity (which is not).

DIRTY DESKTOP
It takes about a quarter-tonne of fossil fuel and one-and-a-half tonnes of water just to manufacture one desktop computer.

And so, in the USA, data centres have sprung up in North Carolina and Washington. These states had surplus electricity capacity, thanks to the collapse of energy-hungry textile, furniture and aluminium industries. So that's where companies such as Apple, Google and Facebook first set up some of their data centres. These are huge, noisy, nondescript buildings full of computer processors, costing up to US$1 billion to set

@DoctorKarl **How come air particles can escape Gravity and not get pulled to the ground?**

up, and chewing up power at up to 100 megawatts, day in and day out. (Australia's total generating capacity is 45,000 megawatts.)

By the year 2020, Greenpeace International say the "annual investment in data centre construction will soar to over $220 billion globally, and $50 billion in the US alone".

The US hosts 40 per cent of the world's data centre servers. They consume about 3 per cent of the entire US energy budget.

NETFLIX

Netflix, a major on-line video streaming service, is a good example of how quickly data usage can change. It turns out that online video is currently the biggest consumer of internet data.

In 2012, the use of Netflix was responsible for 33 per cent of all downstream internet traffic in the USA – between the peak hours of 9 pm and midnight. Another 32 per cent of fixed-line data traffic was accounted for by other (less popular) audio- and video-streaming services. From 2011 to 2012, the average monthly wired connection data download had more than doubled from 23 gigabytes to 51 gigbytes.

Netflix entered the Australian market in 2015, and within weeks it accounted for 15 per cent of all Australian internet download traffic.

CLOUD = 2 PER CENT

The energy used by the Cloud is about 2 per cent of the world's energy. If the Cloud were a country, it would be sixth in the world in terms of energy consumption – after the USA, China, Russia, India and Japan, but ahead of Germany.

By itself, Google uses more power than the country of Turkey.

Energy consumption related to the Cloud is increasing at 12 per cent per year. In the average household, electronic devices (with screens) used about 15 per cent of your electrical power in 2011 – and this is expected to reach 45 per cent by 2030.

They get pulled down by Gravity, BUT there is relatively constant input of power (~1kW/m2 @ equator) from Sun → uplift of air, currents.

HIDDEN EXTERNALITY

In Economics, an "externality" is a cost (or benefit) that affects a party – who chose not to carry or incur that cost (or benefit). A "hidden externality" is one that the community it affects doesn't know about.

In the USA, coal-fired power stations generate carbon dioxide – as well as harmful particles, noxious gases, heavy metals such as mercury,

Suppose that you stream an hour of video each week. The power used to get that 60 minutes of video into and out of the Cloud, and then to your smartphone is more than the power needed to run your refrigerator for a week.

In 2011, India had some 300,000 mobile phone towers – 40 per cent of them in areas without reliable grid-supplied electricity. So the power for these towers came from diesel generators. These towers used 2 billion litres of diesel each year to provide the electricity. That made them the second biggest user of diesel fuel in India, after the Indian railway system. These towers accounted for more than 2 per cent of India's total greenhouse emissions.

In the USA, in Silicon Valley, Microsoft's Data Centre has backup diesel-powered electrical generators. Microsoft has been one of the largest stationary diesel polluters in the whole region.

Bitcoin (see page 129) was specifically set up to use lots of power. (This was to change from "trust in people or an institution" to "trust in mathematics".) As a result, the average power consumption of the Bitcoin Network sits between the power consumptions of Iceland and Ireland.

ENVIRONMENTAL DAMAGE

If high electricity consumption is not bad enough, here's another environmental factor – water usage, to a level that competes with both global warming, etc. There are economic, health and environmental costs associated with each of the stages in the life cycle of coal – extraction, transportation, processing and combustion.

The annual total of these individual costs is between one third and one half of a trillion dollars. This expense is carried by the general public. This price is not factored into the cost of generating electricity with coal. So this cost is a hidden externality. The coal electricity industry have the option of paying for it – but they choose not to.

agriculture and humans. In Utah, the new data centre of the National Security Agency (NSA) uses 6400 tonnes of water each day for cooling. (The average domestic use is $1/10$ tonne of water per person per day.)

What about raw materials? The tin that comes from Indonesia is collected by a process that wreaks havoc on humans, animal and plants. Cobalt from the Congo comes via a process akin to slave labour.

In addition, you end up with electronic waste, or "ewaste".

Western nations generate more that 25 million tonnes of ewaste each year. Most of it is exported to Asia where the valuable raw materials are recycled (which is good), using hazardous techniques (which is bad).

But even in Silicon Valley in the USA, where you would expect environmental standards to be higher, there are about two dozen very heavily polluted sites full of extremely hazardous ewaste. Superfund, a US Government fund, has spent over US$200 million trying to clean it up – but that's barely scratched the surface.

TELEPHONE ENERGY

In 2007, the energy footprint for the global telecommunications network was 293 billion kilowatt hours. That was more that the entire electricity demand for Spain (276 billion kilowatt hours) for the same period.

Slightly. Sun burns 620 MTonnes H/sec → 616 MTonnes + 1 MTonnes/sec Solar Wind + 3 MTonnes "lost"/sec (via $E=Mc^2$).

CLEAN AND GREEN?

I love the access to information, and ease of communication, that our high-technology society can give us. But high-tech grew up so quickly that in many cases, it evolved in a messy and dirty way.

China is trying to go green. In 2014, China invested US$90 billion in renewable energy – a 32 per cent increase over 2013.

Some of the big high-tech companies are specifically going out of their way to become clean and green, some are paying lip service (that is, talking the talk, but not walking the walk), and others are deliberately trying to stay cheap and dirty.

But "cheap" is an illusion – it's cheap only for the polluting company. The rest of society has to pay for the clean-up and the health costs – often for decades after the polluting company has conveniently declared bankruptcy.

High technology is here to stay. And it does look so slick and clean, as it brings magic to your screen. But let's make it green behind the scenes – and give that Cloud a silver lining.

"INFORMATION WANTS TO BE FREE"

In the early days of the internet, people often quoted the mantra, "Information wants to be free".

The first written record we have of this mantra dates back to the very first Hackers Conference, in Sausalito, California, in 1984. Steve Wozniak, co-founder of Apple, had just made an interesting point to the 125 programmers who had turned up to the conference. Wozniak had said that "it was a shame companies wouldn't give engineers the rights to products they developed if the company decided not to market them".

Stewart Brand – editor of the Whole Earth Catalog, founder of The Well and The Long Now Foundation, futurist and all-round ultra-cool dude – then responded to Steve Wozniak. He said, "On the one hand, information wants to be expensive, because it's so valuable. The right information in the right place just changes your life. On the other hand, information wants to be free, because the cost of getting it out is getting lower and lower all the time. So you have these two fighting against each other."

Quick as a flash, realising he had just been exposed to very deep concepts, Wozniak said, "Information should be free, but your time should not."

A compost pile puts out thousands of times more power (weight for weight, and volume for volume) than the Sun.

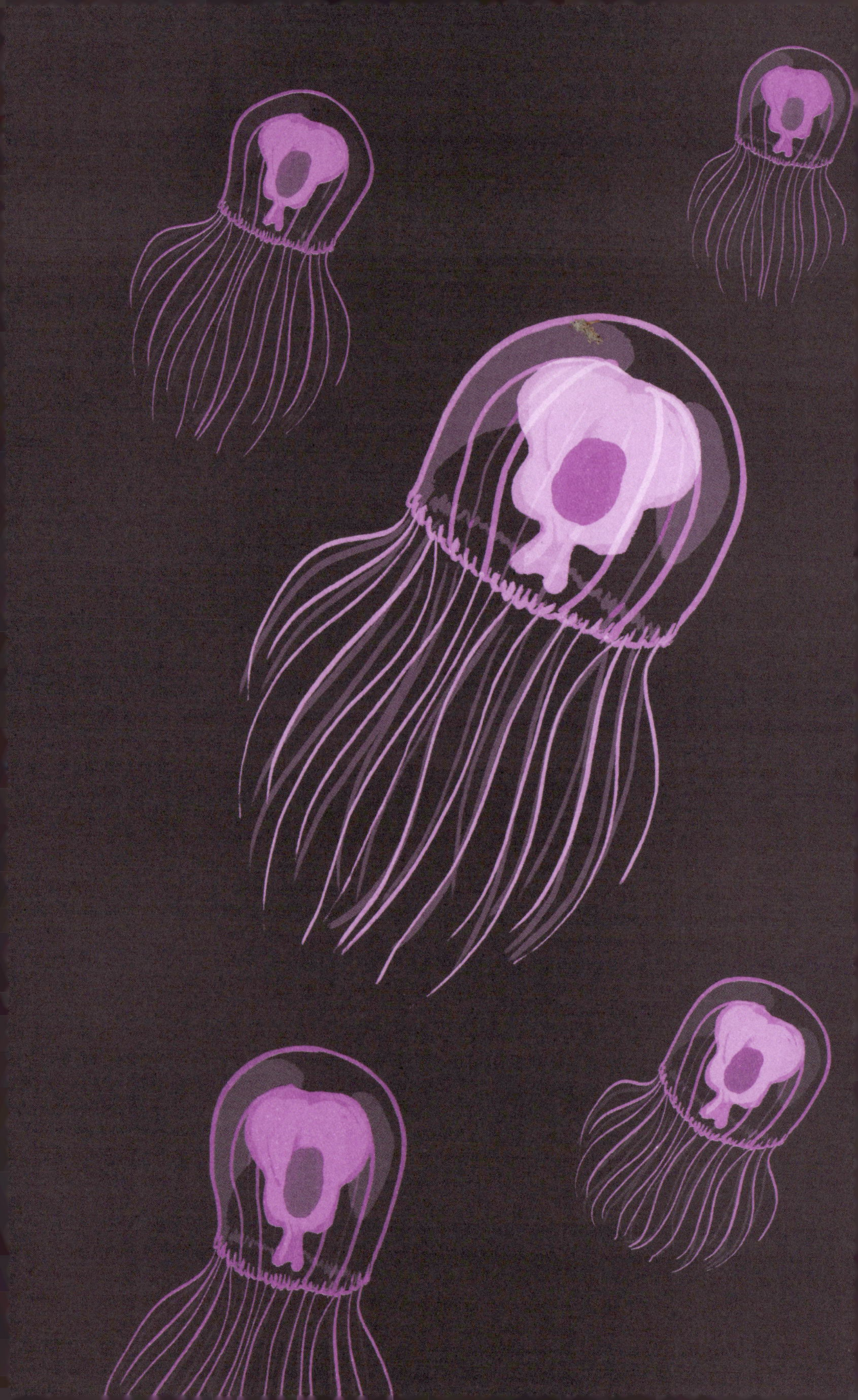

18

IMMORTAL JELLYFISH

DOWN THROUGH THE AGES, THERE HAVE ALWAYS BEEN MYTHS ABOUT IMMORTALITY – BEING ABLE TO LIVE FOREVER. FINALLY, OUR MARINE BIOLOGISTS HAVE FOUND A CREATURE THAT COMES CLOSE: A TINY TRANSPARENT JELLYFISH.

JELLYFISH 101

Jellyfish are special – they have neither a brain, nor a heart. They have only a single opening through which food comes in and waste comes out. So jellyfish eat via their anus.

Jellyfish are the most efficient swimmers in the oceans. They use less energy to cover a given distance than any other ocean creature.

How do they gain that extra efficiency?

Well, first, they push water directly away from the direction they want to travel in. That motion is more efficient than pushing water to the side like fish.

Second, when their umbrella-shaped bell contracts, it creates two vortices, or rotating rings of water. The first vortex pushes away from the jellyfish. The second begins to spin, which sucks in water and gives a free push to the jellyfish. So the jellyfish can travel an extra 30 per cent for no extra energy.

SPREADING WORLDWIDE

T. dohrnii are spreading through the oceans of the world. It seems as though they are getting free rides in ships. When in port, after unloading their cargo, ships suck in water to fill their ballast tanks so they ride better at sea. This water is sometimes loaded with *T. dohrnii* ready for an adventure.

IMMORTAL JELLYFISH

While we unofficially call it "Immortal Jellyfish", its official name is *Turritopsis dohrnii*.

After an adult male *T. dohrnii* squirts his sperm into the ocean waters, some of them end up inside a female *T. dohrnii*. This creates fertilised eggs, which then turn into tiny free-swimming larvae called "planula". After a while, the planula give up swimming, dive down to the sea floor and attach themselves to a rock.

They then change shape entirely into columns of highly branched polyps.

After a few days, another shape change happens. Tiny jellyfish (about 1 millimetre across) "bud" off from the tips of the polyp and, like miniature umbrellas with tentacles, float through the ocean. After 2 to 4 weeks, they become sexually mature males or females. They're now about 5 millimetres across, and their bright red stomach is visible through their transparent body. They eat plankton, tiny molluscs, larvae and fish eggs.

Are *T. dohrnii* jellyfish impervious to all threats? No, they can be eaten by bigger creatures, or, say, get sucked into a vent of a nuclear power plant – so they are not unkillable.

But the "immortality" steps in when *T. dohrnii* suffers an attack, or starvation, or some other kind of environmental stress. Instead of dying, it changes first into a tiny blob, and then within three days reverts back to the polyp stage. It regroups as a polyp colony sitting on a rock. This new polyp is genetically identical to the original jellyfish – but is packaged differently.

Did that original jellyfish die? Not really.

Did that original jellyfish continue to live in the same body? No. It's kind of like a butterfly that, instead of dying, changes back to a caterpillar – or an aged chicken turning back into an egg. This is not a blueprint for humans to use, so that we can cycle indefinitely between a baby and an aged adult.

Technically it's more like "Regeneration" – but it's the closest that we know of to immortality. And once we learn how *T. dohrnii* does it, we could apply this knowledge to medical science for humans.

MORE THAN ONE IMMORTAL?

It seems that there may be more than one species of immortal jellyfish. We are not entirely sure, but it seems very likely that the jellyfish *Laodicea undulata* and *Aurelia* can also change back to the polyp stage.

Visual illusion. Take spirit level with you.
Many such "Anti-gravity Hills" around world.

19 HOT TEA COOLS YOU DOWN

IT SEEMS IMPOSSIBLE THAT SIPPING A NICE HOT CUP OF TEA ON A HOT DAY WOULD COOL YOU DOWN. BUT, UNDER CERTAIN CIRCUMSTANCES, IT CAN – THANKS TO SWEAT.

It takes energy – heat – to make sweat, and it takes lots more energy to make it evaporate.

I first heard (and wrongly disagreed with) the cooling effect of tea while studying Physics. I thought adding extra heat from the steaming cuppa would simply make things hotter.

I didn't understand sweat. Back then, I had no idea of how the body worked – it would be at least 10 years before I learnt any Physiology.

When you drink hot tea, it can trigger temperature sensors in your mouth that activate sweating. (These are TRPV1 receptors. They also get activated by chilli, as in curry.) Tiny droplets of sweat exude from your skin all over your body – which takes energy. The surrounding hot air evaporates the sweat – and, thanks to the Laws of Physics, this takes lots of energy. The droplets of sweat sit on your skin, so your skin cools as they evaporate.

@DoctorKarl **Why can't iron gets pulled out of your blood with centrifuging, like plasma get separated from the blood?**

But evaporation won't work on a very humid day. When there's a lot of water vapour already in the air, the water in the sweat on your skin won't evaporate easily. It will probably just soak into your clothes, or drip to the ground.

Of course, the effectiveness of a hot cuppa also depends on how hot your drink is, how much you drink, the temperature of the day, if you're exercising, and more.

So on a hot and humid day, if you're exercising hard and generating more sweat than evaporation can remove, you're probably better off with a cold drink.

But, under the right circumstances, a nice cuppa can cool you down. And so might a hot curry.

1) Iron is bound to various transport chemicals, hard to separate
2) Plasma (55% of blood, other 45% is cells) is just salty water → easy.

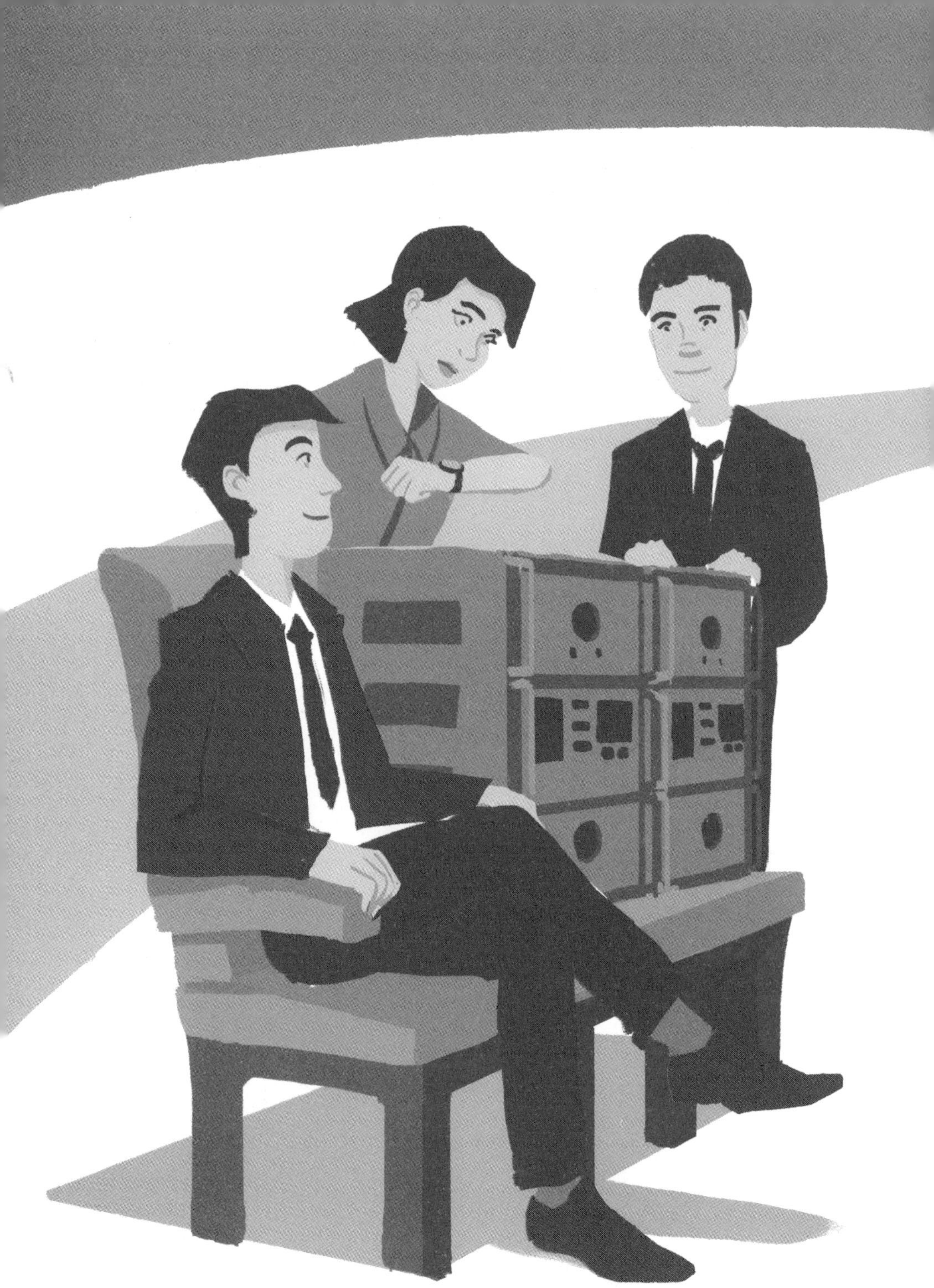

20

TIME TRAVEL

THERE ARE TWO TYPES OF TIME TRAVEL – INTO THE FUTURE, AND INTO THE PAST. PAST TIME TRAVEL IS PROBABLY IMPOSSIBLE, WHILE WE ALREADY TRAVEL INTO THE FUTURE ALL THE TIME.

FUTURE TIME TRAVEL

On average, we are all travelling smoothly into the future at the rate of one second per second. But Einstein's Theories of Special and General Relativity have given us two ways to slow this flow of time down – speed and gravity. Both high speed and high gravity will slow the flow of time.

One of the earliest (and cheapest!) experiments was carried out in October 1971. It took more than half a century after Einstein published his

Theories in 1905 and 1915 for everything to come together. Atomic clocks had become precise enough to get sufficiently accurate measurements, as well as being rugged and portable. Also, thanks to the recently introduced Boeing 747, jet travel was both fast enough that the "slowing" of time was measurable and cheap enough for the scientists to afford it.

Joseph C. Hafele (a physicist) and Richard E. Keating (an astronomer) took four caesium-beam atomic clocks on round-the-world trips. They booked four seats – two for themselves, and two for Mr Clock (with two atomic clocks in each seat).

The total cost was US$8000 – with over US$7600 being the price for eight round-the-world tickets. The six of them (two humans and four atomic clocks) flew around the world twice: eastward, and then westward. Their results agreed "reasonably" with Relativity – showing that time slowed down with increased speed. Later re-enactments of their initial experiment with better atomic clocks gave much better agreement with theory. But jet planes are slow compared to light.

A spaceship travelling at 90 per cent of the speed of light (270,000 kilometres per second) would have its time slow down by a factor of 2.6.

What is the fastest matter propelled regularly by human technology? At the moment, it's the hydrogen ions (H^+) zipping through the 27-kilometre circumference of the Large Hadron Collider. These hydrogen ions (also known as protons) travel at 99.999 999 1 per cent of the speed of light. Their time slows down by a factor of 27,777,778. One second for one of these protons is about 11 months for us.

Gravity also slows down time. Imagine two atomic clocks on shelves, the upper one being 33 centimetres above the lower one. The upper one experiences slightly less Gravity, because it is just a little further from the centre of the Earth. In 2010, our technology was precise enough that we could measure that the lower clock (in the slightly higher gravitational field) was ticking more slowly.

Of course, when we talk powerful gravitational fields, we think of black holes. There is an enormous one at the centre of our galaxy, but there's a much smaller one 3000 light years away. The movie *Interstellar* has a planet orbiting very closely to a black hole. In the movie, one hour spent on the planet has the people back on Earth ageing seven years.

PAST (AND FUTURE) TIME TRAVEL

The famous physicist Stephen Hawking hosted a party for time travellers in 2009. Then in 2010 he sent out the invitations. Of course, nobody showed up. Does this prove that time travel into the past is impossible? Of course not – it's just one experiment.

There are a few different "theoretical" ways to travel to the past. But none of them will take you back to before your Time Machine was built.

One theoretical way to travel to the past involves wormholes or Einstein–Rosen bridges. Let me emphasise that so far we have zero proof of the existence of wormholes. But wormholes are perfectly reasonable theoretical constructs in the Land of Physics. It's been theoretically proposed that wormholes could be a pathway out of a black hole.

Let's pretend they are real, and that we can build one. Assume that our wormhole has both ends in our laboratory. A wormhole can be very stretchy – you can pull on it all you like, and it won't break.

So nail down one end of the wormhole (A) so that it does not move. Then, take the other end (B) for a drive at very, very close to the speed of light for an hour, before returning to the laboratory. Time has moved normally (at one hour per hour) for End A – but has practically stopped for End B. End A thinks the time is 11 a.m., but End B thinks the time is 10 a.m. (They're both right, but that's Special Relativity for you.)

Now jump into the wormhole. (Ignore the inconvenient fact that the diameter of a wormhole is probably trillions of times smaller than an atom.) If you jump into End A and pop out at End B, you will have gone one hour into the past. And vice versa – jump into End B, and when you emerge at End A, you will have arrived one hour into the future.

By the way, it isn't just you having trouble with getting your head around "wibbly-wobbly timey-wimey stuff". Entering the Land of Theory can do very strange things to your Sense of Equilibrium.

Depends on time, moisture level, temperature, local bacteria. Smell is usually a good test.

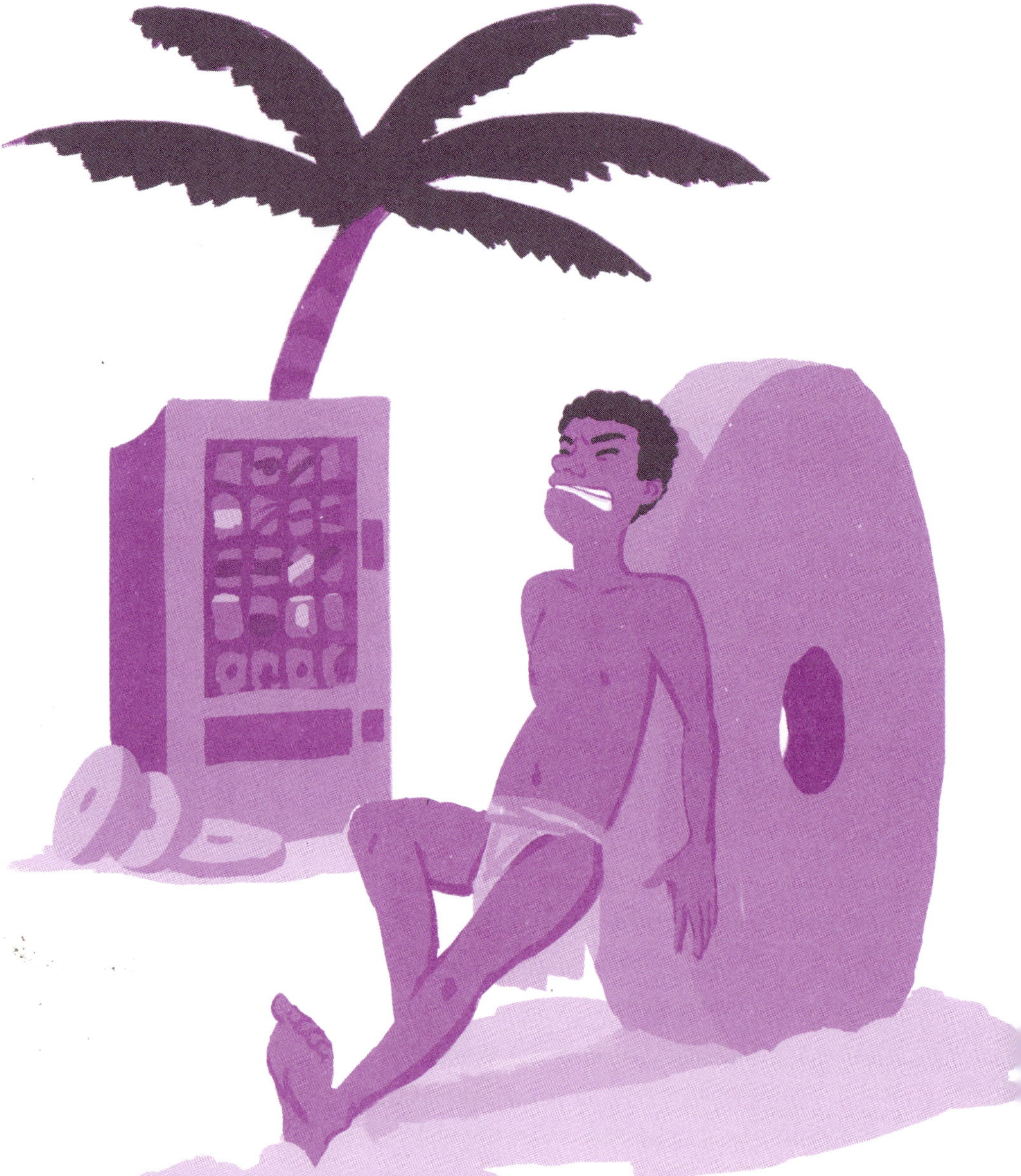

21

BITCOIN: LEGEND OF A LEDGER

TODAY, PRACTICALLY ALL SOCIETIES USE SOME KIND OF MONEY AS A CONVENIENT WAY OF EXCHANGING GOODS AND SERVICES.

Down the ages, money has been through numerous physical incarnations: promises scratched into soft clay that when fired became unchangeable, sea shells, large wheels of limestone rock with a hole in the middle, metal coins, paper notes, cheque books, credit cards and so on.

In each case, "money" involved some kind of physical object. But now "virtual money" has arrived.

And its first commercial use was to buy two pizzas, for the princely sum of 10,000 bitcoins – which later rose in value to many millions of dollars.

TRUST IN YAP

Trust is essential to every currency. Currency has value for two reasons. First, the State has willed the currency into existence. Second, the citizens trust that fellow citizens of the State will accept the currency from them.

As an example of trust, the people of Yap have so much faith in their currency that it can sit, unreachable, on the ocean floor – and still be accepted as valid.

The four Yap islands sit inside a coral reef, and carry a population of 11,000 people, spread over some 100 square kilometres. These islands are about 1200 kilometres north of New Guinea and 1200 kilometres east of the Philippines – and have no native metals or ores.

So for over a thousand years (until very recently), their currency was based on Stone Coins, which they call "rai" (the word "fei" is also used). The rai are fashioned from shiny limestone rock available only on an island called Palau, some 400 kilometres away. The Yapese sail over, quarry the blocks of limestone (without metal!), and fashion them into wheels with a hole in the middle. They load the rai onto a raft and tow them back to their home islands of Yap, to use as money.

These Stone Coins range in size from a clenched fist up to 3.6 metres across, and can weigh up to four tonnes. They hold their value well, because they are difficult to get, and quite hard to counterfeit. Each stone's actual value is complex – it depends on its size, how hard it was to quarry and bring back, how many were injured or died in getting it home, if somebody famous was involved, and a whole host of other factors.

You might have a big Stone Coin sitting outside your house – and one day, decide to spend it all to buy several pigs from a merchant. You and the merchant come to an agreement, and pretty soon, everybody knows that you swapped your Stone Coin for some pigs.

Now here's the weird part. The Coin doesn't shift – it's still outside your house, but everybody knows it now belongs to the merchant.

@DoctorKarl **How and why does the microwave interfere with my Bluetooth?**

And if the merchant then sells that specific rai to buy some more pigs, the seller of the pigs will probably leave the Coin outside your house. Rai are big and heavy, and hard to shift, and you don't want to damage them.

So in Yap, the "money" is actually the community memory – which in turn, is the collective Register of Ownership.

Now here's something even weirder. Sometimes, as the Yapese return from Palau with their hard-won rai, they get hit by a big storm. They have to jettison their huge Coin so their tiny boats won't sink. But it took a lot of work to make that Coin. So when they return safely, they tell everybody what happened, and everybody trusts them and then assigns the ownership of that rai to the person who quarried it.

Even though that specific Coin is sitting many kilometres down on the bottom of the Pacific Ocean, it can still be traded as legal tender. Why? Because the people in Yap society have trust in their currency, and in the register of transactions – which in this case, is the collective memory of the citizens of Yap.

TRUST IN RECORDS

In the West, we have a very different system from the one in Yap – but it's still built on trust.

It probably began in ancient Sumer (modern-day Southern Iraq), about 5000 years ago. The Sumerians kept written records of what they traded and received.

A more recent example (1397–1494) was the Medici Bank in Florence. Thanks to the popularisation of the balance sheet, the Medici Bank became the place where all the different credits and debts of the citizens of Florence were gathered into a single register.

This system worked because the citizens of Florence had trust in the integrity of the Medici Bank.

Bluetooth (2.4–2.485 GHz) & microwave oven (2.45 GHz) operate on electromagnetic frequencies that overlap → interfere.

TRUSTING 90 PER CENT

Back in 2006, the sum total of all the money on Planet Earth was around US$473 trillion. But less than 10 per cent of it was actually physical money – banknotes and coins. Over 90 per cent was simply entries in a ledger (a book, or collection, of financial accounts). The vast majority of that 90 per cent was stored electronically, with no paper record.

BITCOIN HISTORY

Back in 2008, many were getting disillusioned with "conventional" finance. This was exacerbated by the US Sub-Prime Mortgage Crisis, which led to the infamous Global Financial Crisis. So some people thought the time was right for a virtual electronic currency that was independent of banks and governments – and more importantly, did not need either banks or governments to be the "trusted" clearing house.

The first workable digital money, bitcoin, was created in 2008 by the mysterious programmer Satoshi Nakamoto. In a landmark self-published paper, he/she/they described how to set up a decentralised and yet secure digital cash system. (For the sake of simplicity, I'll refer to Satoshi as female.) She came up with the brilliant concept of the "blockchain". The blockchain is both a record of the transaction, and a solution to the fundamental problem of "Double Spending".

With a banknote, you can spend it only once. This is because in "regular" currencies, only the trusted central bank is allowed to print money. Citizens aren't allowed to manufacture the physical currency.

But in a computer, Double Spending is as simple as cutting-and-pasting some text as often as you like. If there are no safeguards, a crook can spend the same bit of digital currency hundreds of times.

A possible digital solution could be to have a trusted third party keep a record. Most of us are cool with this. Today, as Nakamoto wrote,

> So when we "spend" money, we simply change entries on an electronic ledger. Once again, the market is built on trust.

"commerce on the Internet has come to rely almost exclusively on financial institutions serving as trusted third parties to process electronic payments". But what if this honest third party started off straight, but later went crooked? (Remember the Global Financial Crisis?)

A bitcoin is not a physical thing – it is just an entry in an electronic blockchain (a set of bytes, or 1s and 0s). The Bitcoin Network uses peer-to-peer networking combined with fancy encryption mathematics to both generate and exchange these bitcoins.

The blockchain is an up-to-date, open, transparent and unchangeable public ledger of all past transactions. The blockchain stops Double Spending. It's a continuous series of blocks of transactions, with a new block added every 10 minutes (or 2016 blocks every two weeks). It takes a lot of work to add each new block – and as it's added, that block is checked for accuracy.

The blockchain goes right back to the very first block that Nakamoto created in 2009.

It doesn't rely on trusting a single central authority (such as your government) or a group of people (such as the entire community of people in Yap).

No, the blockchain relies on strange One-Way Mathematics that is easy to do, but hard to undo – just like a big Stone Coin.

Tides act on rotating Earth ➜ Earth rotates slower ➜ Moon moves away @ 4m/century to balance angular momentum equations.

ONE-WAY MATHS AND SECURITY

You might think that the concept of One-Way Mathematics is ridiculous. After all, you can start with 3, then add 2 to get 5. Then you can use subtraction to get back to where you started. Yes, 5–2 gets you back to 3. This is Two-Way Mathematics.

But think about multiplying two really large prime numbers, each 500 digits long. It would take many hours by hand, or a few instants on a computer. The multiplication gives you a really long number, about 1000 digits long.

Welcome to One-Way Mathematics.

> 457 x 1,299,709 = 593,967,013
>
> Multiplying a 3-digit prime and a 7-digit prime to get a 9-digit number is easy, but finding those two primes from the 9-digit number is very hard.

It's practically impossible to go backwards and find the only two prime numbers that can be multiplied to generate this 1000-digit number. You just have to make lots of guesses, and try them out – one after the other. This is appropriately called the "Brute Force Method". You might be lucky and guess the correct answer on the first attempt – or you might not.

It's not totally impossible to find those two prime numbers – but it would take today's fastest supercomputers many times the age of the Universe to find them. So with today's technology, it's effectively (or computationally) impossible to find the only two factors of our 1000 digit number.

One-Way Mathematics underlies the two key technologies needed for the blockchain to work. The blockchain is the brilliant invention that stops people from spending their virtual currency more than once.

The first technology is called Public and Private Key, or Public Key Infrastructure (PKI). It means that if you claim that you bought and sold some bitcoins, everybody can be confident that it was definitely

---

@DoctorKarl Why fly no fry in microwave? (i.e., how come insects survive inside operating microwave oven?)

you (and nobody else) that made this claim. It provides secure communication that nobody else can intercept or alter.

The second technology is the Hash Function. In this case, the Hash Function is used specifically to make it really hard to add the latest block of transactions to the existing blockchain. (There's a reason it's hard.) Every active member in the entire Bitcoin Network receives the latest block of transactions at about the same time. They all try to add that block to the existing blockchain. On average, only one member will succeed, and on average, it will take them about 10 minutes. It takes a lot of work from the entire Bitcoin Network for one person to add that latest block.

If you wanted to add a false block to the existing blockchain, you would need more computing power than the entire Bitcoin Network. And how much power is that? Well, it varies from month to month, but the massed computing power of the Bitcoin Network is between 250 and 10,000 times the computing power of the top 500 most powerful supercomputers in the world combined.

So the Hash Function effectively makes the blockchain "honest". The integrity of the blockchain is based on the fact that it takes a lot of work to add each new block. (But with access to unlimited computing power, you could undermine the blockchain.)

PUBLIC AND PRIVATE KEY

One-Way Mathematics is essential to the strange encryption method called "Private and Public Keys", or "Public Key Infrastructure". It was proven to be theoretically possible in 1976 by Whitfield Diffie and Martin Hellman. This set off a race among mathematicians to make it work. The first Public and Private Key Encryption was actually constructed two years later in 1978 by Ronald Rivest, Adi Shamir and Leonard Adleman, later collectively known as RSA. You might have read of Julian Assange and others using this technology to conduct totally secure communications.

Here's how it works.

Wavelength of microwaves in oven ~10 cm → hotspots 10 cm apart → blowfly avoids heat and moves to cold spot → lives.

In cryptography, the examples always start with Alice and Bob. If there's an Eve, she's the eavesdropper. "Cryptography" comes from Greek words meaning "secret writing".

Alice mathematically constructs two keys, as a related pair – a Private Key and a Public Key. (Each key is just a bunch of letters and numbers.) Everybody in the world can have a copy of Alice's Public Key. They can plug Alice's Public Key into the RSA Algorithm (also freely available to anybody who wants it) when they want to send her a message.

So Bob uses Alice's Public Key to encode a message and email the resulting string of gibberish to Alice.

Alice is the only person who has her Private Key. She applies her secret Private Key to the gibberish Bob emailed her – and suddenly, Alice can read Bob's original message. This is the weird bit – even though everybody has Alice's Public Key and a copy of the RSA algorithm, she's the only person who can decode it.

But besides making it possible to send messages that are totally secret, Public-Key Cryptography makes it possible for each of us to have a unique Digital Signature.

A Digital Signature is like your regular signature – easy to make, hard to forge. In fact, it's so hard as to be computationally impossible. Again, it's a two-part process. One algorithm is used with your Private Key to sign the message, and another algorithm is used with your matching Public Key to check the validity of the message.

You can take it one step further.

You can fuse your Digital Signature to your message so they can't be separated. This means your Digital Signature can't be copied and used on another message. So anybody can verify that the message they received came from you – and only you.

The Digital Signature provides proof of ownership of bitcoins. This is how human and institutional trust can be replaced by mathematical trust.

It means you can proclaim to everybody that you promise to send me 10,000 bitcoins – and everybody can be confident that you are the person who made that transaction, and actually own those bitcoins. (However, "you" in the Bitcoin Network is just a bunch of numbers and letters. I have no idea who you really are, and where you live. But thanks for the 10,000 bitcoins . . .)

By the way, if you want to look at the continuous flow of transactions on the Bitcoin Network, check out https://blockchain.info.

Bitcoin: Legend of a Ledger **< 137**

MSG: HELLO ALICE!

Bob uses Alice's public key to encrypt

AF829GC1DA3715BC sent to Alice

Alice uses her private key to decrypt

MSG: HELLO ALICE!

Pathway/method to have fuel for mobile vehicles (with fuel that is NOT carbon-based). Also need H Fusion for Space Travel.

HASH FUNCTION

The second thing involved in understanding the blockchain is the "Hash Function". This is nothing to do with hashtag (as on Twitter), or hashish (the drug). It gets its name from cooking, as "hash" meaning to "chop and mix". A Hash Function will chop the input and then mix it up, following a whole bunch of crazy mathematical rules, to give an output. The Hash Function that the Bitcoin Network currently uses is called "SHA-256". The output, a unique chain of letters and numbers is called a "Hash Value", or a "Hash Code", or a "Hash Sum", or simply, "hash".

First, the same input always gives the same output (or hash).

Second, two different inputs can't generate the same hash. (Well, hardly ever. It's very very rare. If it happens, it's called a "collision".)

Third, you can feed into a Hash Function any number of characters – a single letter or number, or the 44 million words in the *Encyclopaedia Britannica*. Regardless, you will always get a hash with the same amount of letters and numbers – 64 of them – in SHA-256. Try it out here: http://www.xorbin.com/tools/sha256-hash-calculator. The letter "c" turns into the hash "2e7d2c03a9507ae265ecf5b5356885a5 3393a2029d241394997265a1a25aefc6". Each time you apply SHA-256 to the Encyclopaedia Britannica, you will get exactly the same hash. But change one single letter in just one of those 44 million words, and the resulting hash is completely different.

And like our 1000 digit number with only two factors, you cannot go backwards. When you look at a hash, you have no idea if the input was a single letter, or the entire *Encyclopaedia Britannica*.

The only way to solve a Hash Function is the Brute Force Method. Guess some characters, see if they give you the right answer, and if they don't, repeat – millions, billions and trillions of guesses. This happens around the world in the 10-minute refresh cycle of the bitcoin ledger (blockchain).

BLOCKCHAIN 101

In a transaction, you might buy a lawnmower for a certain number of bitcoins. But for your transaction to be valid, it has to show from whom you got those bitcoins, and to whom you are sending them. It also has

---

@DoctorKarl What gets left after burning Hydrogen? (eg, on Earth, and in Space).

to incorporate a timestamp. Transactions are grouped into 10-minute long groups called "blocks".

These blocks are broadcast to every "node" (networks of super-fast computer processors) in the Bitcoin Network. Each node tries to add the new block to the last block in the existing blockchain – and if it's the first to succeed, it gets paid in bitcoins. This is what they call "mining" bitcoins.

The name "miner" comes from "gold mining". Suppose you discover some gold in your backyard. It will cost you money to mine that gold, and after a while, all the gold will have been mined.

Nakamoto herself mined the very first block of 50 bitcoins on 3 January 2009. Part of her Grand Plan was that the payment would halve roughly every four years (or 210,000 transactions). By 2013, the number of bitcoins in a payment to the miner had halved to 25. In mid-2016, the number of bitcoins in circulation was around 15 million. The mining payments for verifying the blockchain will drop to zero around 2140, when some 21 million bitcoins will have been issued. After that, no more new bitcoins will be issued.

After 2140, bitcoin miners will get paid in terms of transaction fees out of existing bitcoins, rather than "mining" new bitcoins out of the virtual ground.

But as the node adds that new block of transactions to the existing blockchain, it has to mathematically incorporate all the information that existed in the previous block at the end of the blockchain. (It combines the hash of the new block with the hash of the previous block.)

Now that previous block incorporates all the information in the block just before it – and so on, and so on, all the way back to 2009. So the blockchain is *not* a set of individual and isolated blocks of transactions – instead, it's a chain of blocks, each of which incorporates all the information in the block immediately before it.

The blocks are tied together in a mathematical cryptographic chain.

1) Burn Hydrogen with Oxygen → Water
2) Burn H in fusion reactor (e.g. small scale Sun) → He.

BITCOIN 404: MORE POWERFUL THAN THE TOP 500 SUPERCOMPUTERS!

For a miner to add the latest 10-minute transaction to the blockchain, they have to solve a very specific Mathematical Problem. To earn their payment of bitcoins, they have to be the first to guess a hash called the "nonce".

(In English Grammar, a nonce is a word created to solve a problem in communication. For example, a "fluddle" is a spillage of water, smaller than a "flood", but bigger than a "puddle". In Cryptography, a "nonce" is an arbitrary number that is used only once.)

You start by combining the hash of the most recent block of transactions with the hash of the last block in the existing blockchain. Apply the Hash Function to that combination to get a third hash. (Remember, every hash is 64 characters of "gibberish" letters and numbers.) You then combine that third hash with the hash that is the "nonce". There is no mathematical way to work out the nonce. You have to guess a value for the nonce, plug it in and try – and repeat until you get the "right answer". It's a lottery. You might be lucky with your first guess, or not.

And what is the "right answer"? Another hash with a certain number of leading zeros – e.g. 0000001e8d3a.

The bitcoin software "wants" the time to solve the problem to be about 10 minutes. It automatically makes the Mathematical Problem more difficult (when it begins to take less than 10 minutes to solve) by adding another leading zero.

This computing problem is then called the "proof-of-work".

By the way, the only way that the blockchain can be changed is by doing the proof-of-work. The difficulty is that if you're trying to change a transaction that happened 50 minutes ago, everybody on the Bitcoin Network has a copy of the blockchain up to 10 minutes ago. In general, the longest chain that arrives at your computer on the Bitcoin Network is accepted as the True Chain. This is seen as proof that it came from the largest pool of computer power.

Nakamoto wrote, "The network timestamps transactions by hashing them into an ongoing chain of hash-based proof-of-work, forming a record that cannot be changed without redoing the proof-of-work. The longest chain not only serves as proof of the sequence of events witnessed, but proof that it came from the largest pool of CPU (Central Processing Unit) power."

One way to change the blockchain is by having more computer power. At the moment, nobody has. In fact, the combined computing power of all the bitcoin computers running at any given instant is at least 250 times greater than the combined power of the top 500 supercomputers in the world!

This is the greatest amount of computer effort ever devoted to a single problem. What could we find out if this computational power was used to solve protein folding problems in drug design, look for anomalies in astronomical observations or analyse any medical problem you can think of?

BITCOIN MINERS

Miners on the blockchain use fancy computer processors and lots of electricity to mine bitcoins. They have to solve a problem. Actually, "solve" is the wrong word – "guess a solution" is more accurate. They keep making guesses until they guess correctly – it's more like a lottery. But everybody calls it a "problem", so I'll go along with that.

Without miners, there is no updating of the blockchain every 10 minutes. Miners are essential to blockchain integrity.

The software is set up so that the first person to solve the problem automatically adds the latest transaction to the chain of all previous blocks (of transactions). They then broadcast the new, updated blockchain to everybody in the Bitcoin Network. The other nodes check the validity of the hash function. If the latest update is valid, they show they accept that solution by incorporating it into their version of the blockchain. As soon as that solution is announced, every other miner in the Bitcoin Network immediately stops trying to solve the previous problem, and moves onto the next block of 10 minutes of transactions. (Of course, the other miners accept the updated blockchain only if all transactions in it are valid, and bitcoins haven't already been spent.)

Nakamoto figured that people would volunteer to maintain the blockchain because they would be paid in bitcoins. The miners

---

Everthing disturbs Universe. Sun burns 620 million tonnes H/sec → 616 million tonnes of He + (3 million tonnes → energy) + (1 million tonnes → solar wind).

8000 CURRENCIES IN THE USA

In the 1860s, there were some 8000 private currencies circulating in the USA. They were issued by banks, railroad companies, retail stores and other entities.

The average citizen could hold a dozen different currencies. Unfortunately, while the currency issued by a restaurant could be traded at full face value in that restaurant, it would be worth significantly less in the

would run special software that would use lots of computer time and electricity. It was the miners who would protect the integrity of the blockchain, not a central trusted authority.

So right from the beginning, mining was designed to be arduous. Back on 12 April 2013, the cost in electricity to run the Bitcoin Network for 24 hours was about US$147,000. The cost of the computers/processors was more again. But the profits from mining those bitcoins came to about US$681,000.

BITCOIN BUYS PIZZA

Laszlo Hanyecz, a Florida programmer who was one of the first bitcoin miners, carried out the first real-world bitcoin transaction. He had mined 10,000 bitcoins. Laszlo wanted to turn his 10,000 bitcoins into piping-hot pizza. But the local pizza store did not accept bitcoins. (Of course it didn't – no commercial institution accepted bitcoins. Back then, bitcoins were like Monopoly Money – essentially worthless.)

So on 18 May 2010, Laszlo posted a request on Bitcoin Forum asking if anybody could use his 10,000 bitcoins (then worth about US$41) to buy two large pizzas (then costing about US$25). As Laszlo said, "what I'm aiming for is getting food delivered in exchange for bitcoins where I don't have to order or prepare it myself." It was purely an experiment to see if he could spend bitcoins.

@DoctorKarl My kids freaked out when they saw the International Space Station fly over and I told them people lived on it.

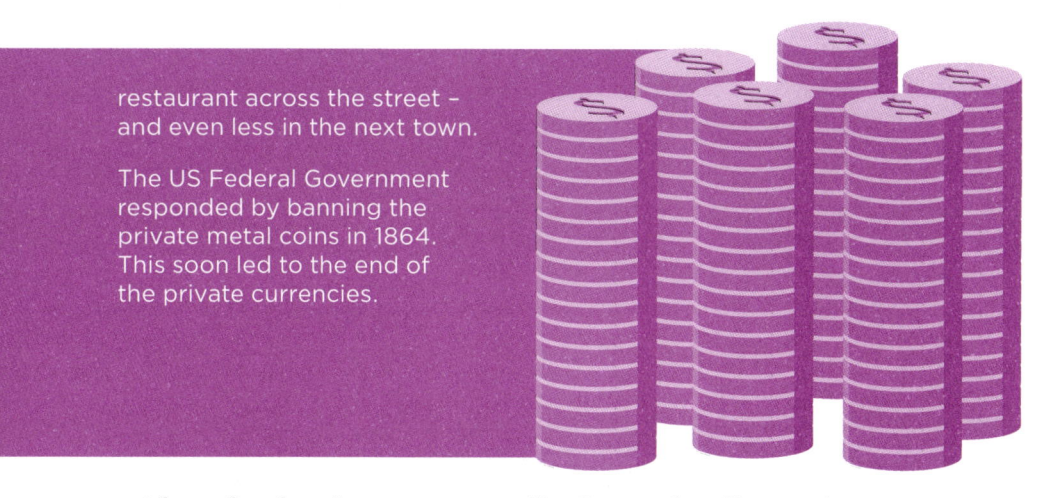

restaurant across the street – and even less in the next town.

The US Federal Government responded by banning the private metal coins in 1864. This soon led to the end of the private currencies.

After a few days, he was contacted by Jeremy Sturdivant, who was then 18 years old, and also living in the USA. After a bit of discussion on Internet Relay Chat (IRC), the deal was finalised. Laszlo sent the 10,000 bitcoins to Jeremy (in other words, he entered this specific transaction into the bitcoin blockchain). Jeremy then contacted a pizza store in Laszlo's city of Jackson, Florida, which delivered the two pizzas. On 22 May 2010, Laszlo joyfully posted, "I just want to report that I successfully traded 10,000 bitcoins for pizza."

By 4 August that year, the value of the bitcoins had risen to US$600, and by 29 November 2010 had reached US$2600. Laszlo said, "I don't feel bad about it. The pizza was really good." By November 2013, those 10,000 bitcoins were worth US$12.4 million. But somebody had to make that first commercial transaction.

To commemorate the importance of that first commercial bitcoin transaction, 22 May is now called Bitcoin Pizza Day. To quote the ancient Chinese proverb, "Give a man a pizza, he'll eat for a day, let him buy pizza with bitcoin, revolutionise the economy…"

BITCOIN VOLATILITY

As Laszlo can testify, the value of a bitcoin has been very volatile, ranging from US$0.03 to over US$1200.

Even now, only about 100,000 merchants worldwide accept bitcoin.

And amazingly, we humans have been living on the International Space Station continuously since 2 November 2000.

FAITH AND FAKERY

In the UK, 3 per cent of all the one-pound coins in circulation are fake, according to the Royal Mint. But British citizens don't automatically think that their coins are worth only 97 pence.

No, they can confidently expect to spend their occasional forged coin for its full nominal value of 100 pence. Mind you, every now and

The advantage of bitcoin over credit card payments is that the merchant doesn't pay a fee to the credit card company. They will accept bitcoin via a third party, but will then usually convert it into their own trusted currency. The problem is that the value of bitcoins is volatile – its price relative to other financial assets varies a lot. So, as an asset, a bitcoin is risky to hold – because without any warning, its value might go up or down. Bitcoin is twice as volatile as gold, and three to four times as volatile as the major currencies.

Volatility can happen as a result of a small economy. (Bitcoin is not a major currency, such as the American dollar.) Because the economy is small, holders of a large number of bitcoins can't quickly sell their bitcoins without depressing the market. Another problem following from the small size of the economy is that any bad press can easily depress the market.

So Satoshi decided there would be a gradual release of bitcoins, with the number released regularly halving. This would hopefully create a self-stabilising economy, with neither too few nor too many bitcoins in circulation. Even so, the bitcoin economy is microscopic on the world scale – equal to rounding errors on a large economy.

So if bitcoin is volatile, what's so good about it?

The answer is the blockchain – but it is expensive to maintain.

then their coins get rejected by a vending machine with a slot for accepting coins. But overwhelmingly, British citizens have faith in their coins, and trust them.

WHERE NEXT?

It takes a huge amount of energy to run the Bitcoin Network. This is the cost of making bitcoin a trusted currency, without having to trust people or institutions. (Remember, bitcoin grew out of the financial shenanigans that led to the Global Financial Crisis of 2008.) Back in 2015, bitcoin miners used enough electricity to power 135,000 homes, or run two Large Hadron Colliders at full power. (That was about four times more electricity than the Bitcoin Network used in April 2013.) Indeed, bitcoin miners have a definite preference for locating their mining machines in cold climates, to reduce their air-conditioning bills.

At a personal level, individual bitcoin miners have used so much electricity running high-end computers that they have been raided by the cops, who thought they were using heat lamps to grow marijuana.

Furthermore, as of mid-2016 the Bitcoin Network seems to have major problems. It can't process more than three transactions per second. The block size needs to be increased to allow more transactions to be processed. But Satoshi Nakamoto passed on control to only a handful of people. (They have control in the sense that they have access to the Source Code of bitcoin, and can change it.) They are in an unpleasant internal "civil war" to stop changes to the block size. Sometimes, the Bitcoin Network slows down so much that payments can take hours to be verified by the miners.

Furthermore, well over half of the miners are now in China (thanks to cheap Application Specific Integrated Circuits [ASICs], and artificially cheap electricity). Mike Hearn, who spent over five years being a bitcoin developer, wrote, "At a recent conference, over 95 per cent of hashing power [that allows the blockchain to work] was controlled by a handful of guys sitting on a single stage." What if they were to stop processing blocks of transactions, and adding to the blockchain?

And finally, a change was made to bitcoin which allows a buyer to take back a payment after buying something.

Overall, bitcoin is, at most, a very marginal aspect of the world's financial system.

On the other hand, financial institutions around the world are looking at using less energy-hungry versions of the blockchain to run transactions between small groups of merchants. In January 2016, the global financial elite at the annual World Economic Forum in Davros, Switzerland waxed enthusiastically and optimistically about future applications of the blockchain. Sir Mark Walport, the Chief Scientific Adviser to the UK Government, wrote that "distributed ledger technologies have the potential to help governments to collect taxes, deliver benefits, issue passports, record land registries, assure the supply chain of goods and generally ensure the integrity of government records and services". The blockchain could also be used to set up tamper-proof voting.

There's a lot here to get your head around. Blockchain mining is not for blockheads – but is a bitcoin ample reward? I like my coins whole.

PRIME NUMBERS

Prime numbers are numbers divisible only by one and themselves. They are the building blocks that all other numbers are made from. The Fundamental Theorem of Arithmetic attributable to Greek mathematician Euclid tells us that every number above one must be either a prime or a product of primes.

So if our 1000 digit number was generated by multiplying two prime numbers, then those are the only primes that could be used to construct that number.

@DoctorKarl Phantom Vibration Syndrome from your phone? How come?

WHERE'S SATOSHI?

Satoshi Nakamoto has vanished.

Satoshi's very deep and insightful paper was published under her name in October 2008. It was in an online journal, which didn't need any proof of identity.

What do we know of Satoshi?

First, Satoshi vanished with a million bitcoins (fair enough, she invented bitcoin). At about US$500 per bitcoin, that's about half a billion dollars.

Second, she made some 500 utterances (emails, forum postings, etc.). They were mostly rather technical discussions of bitcoin's source code. Nakomoto's last post to the Bitcoin Forum was on 12 December 2010.

Third, very few utterances were made between 12 midnight and 6 a.m., East Coast USA time. Maybe that was her time zone, and she was sleeping?

Fourth, the name could be made up – "Sa" from "Samsung", "Toshi" from "Toshiba", "Naka" from "Nakamichi" and "Moto" from "Motorola". In Japanese, "satoshi" can mean "wisdom" or "reason", while "nakamoto" can mean "central source".

Fifth, there is a Californian man called Dorian Prentice Satoshi Nakamoto. He worked on some classified military projects. *Newsweek* tracked him down in 2014, but he said he was not "the" Satoshi Nakamoto. Soon after, the real Satoshi came online (after a silence of several years) to write, "I am not Dorian Nakamoto".

Sixth, in 2015, an Australian man, Craig Wright, claimed he was actually Satoshi Nakamoto. As "proof", he re-used an old Digital Signature used by Satoshi Nakamoto back in 2009. But anybody could have copied this particular signature. It was like producing an Australian banknote (which carries the signature of Glenn Stevens, the Governor of the Reserve Bank of Australia) and claiming that this proved you were Glenn Stevens. Some journalists (who did not understand bitcoin) said they were convinced he was really Satoshi. Other journalists (who did understand bitcoin) said that what Craig did in no way proved he was Satoshi. Shortly after, Craig was raided by the Australian Federal Police on unspecified Tax Matters.

Phantom Vibration Syndrome? 1) Real 2) Dunno 3) Neither phantom, nor vibration, nor syndrome 4) "tactile hallucination".

BETTER MINING MACHINES

In the old days, a high-end desktop computer was powerful enough to mine bitcoins. It could do the proof-of-work necessary to add the latest block of transactions to the already existing blockchain.

The next step happened when the miners realised that there was another, and more powerful, processor inside their computer. Yes, in addition to the standard Central Processing Unit (CPU), there was also the Graphics Processing Unit (GPU) – dedicated only to processing graphics (images). It wasn't as flexible, and couldn't carry out as many different tasks as the CPU – but it was lot faster at what it could do. By a lovely coincidence, it could also solve the Hash Function needed to complete the blockchain. Soon, miners posted photos of dozens of these GPUs humming away – eating electricity, verifying the blockchain, and earning bitcoins.

The third step was the Field Programmable Gate Array (FPGA). An FPGA is not something you put into a computer. Instead, it's a standalone piece of hardware that can be programmed for any specific job via the appropriate software.

In early 2013, the fourth step in this escalating Hardware War was taken. Avalon released the first dedicated mining rigs built from Application Specific Integrated Circuits (ASICs). It took years to design and build these purpose-made chips.

In late 2013, a desktop computer with a high-end i7 processor would take about 263 years to mine a bitcoin. A single fast GPU would take 4.17 years, while a single FPGA would take two years. But a single ASIC could do this in 9.1 days. However, this is still a lot longer than 10 minutes. So the miners had three choices. Pray for luck, buy banks of ASICs, or combine your ASIC with those of other miners – the so-called "mining pool".

Perhaps, just like in the California Gold Rush in 1849, the suppliers of mining equipment will, on average, make the most money.

@DoctorKarl What makes grey hairs grow quicker?

Bitcoin: Legend of a Ledger **< 149**

Hair follicle grows hair faster → make Reactive Oxidative Species → "kill" melanin-producing part of follicle → grey hair.

22 SPLEEN AND RED BLOOD CELLS

IF THE SPLEEN REMOVES OLD RED BLOOD CELLS, WHAT HAPPENS WHEN YOU LOSE YOUR SPLEEN?

SPLEEN 101

The spleen sits in the upper left quadrant of your tummy, hiding behind your lower ribs. It's about 7 to 14 centimetres long, between 150 and 200 grams in weight, and rather soft and pulpy. The White Pulp (that's its real name) does lots of Immune System stuff, while the Red Pulp mechanically filters out old red cells. It also stores about 240 millilitres of red blood cells, which it can release in an emergency (e.g. massive bleeding). If you have a heart attack, the spleen will

@DoctorKarl Is it legit to think of Gravity as the warp, which results when Mass shifts Space?

release huge numbers of monocytes (white blood cells) that will migrate to the inflamed muscle tissue of the heart to remodel the damaged tissue. Because the spleen is so soft and squishy, it's easily damaged by trauma and yet virtually impossible to repair. So accidents can be a reason people need their spleen removed. Luckily, you can survive pretty well without a spleen. However, you'll have reduced responsiveness to certain vaccines, increased susceptibility to certain bacteria and protozoa, and increased chance of diabetes and heart disease.

You make about 2.4 million red blood cells each second. They begin to age when they are around 100 to 120 days old.

So when you spleen is missing, how do damaged and/or aged red blood cells get taken out of circulation?

The body often has back-up systems. After a splenectomy, Kupffer cells that live in the liver swing into action. (Kupffer cells are macrophages, one of the many, many types of immune system cells.) In one study, damaged red blood cells were radioactively labelled (so they could be tracked). When they were injected into mice, over half of the damaged red blood cells were removed within 10 minutes. Half of those were taken out by Kupffer cells.

So you can get away with being spleen-less, but not heart-less...

Yes-ish. Mass tells Space–Time how to curve. Curved Space–Time tells Mass where to go.

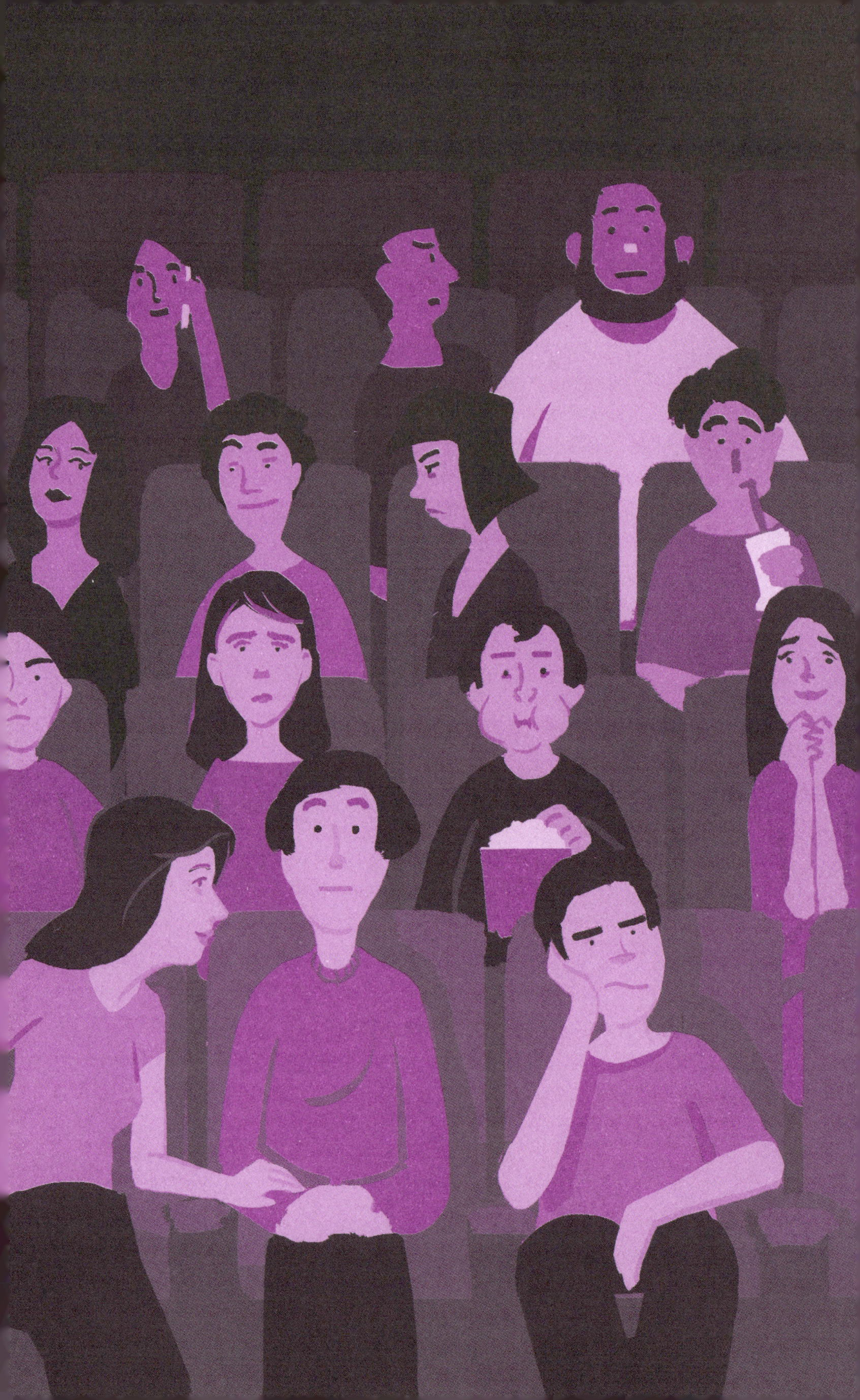

23

MOVIE AUDIENCES EMIT CHEMICALS

NOWADAYS, THANKS TO THE INTERNET, YOU DON'T HAVE TO GO TO THE MOVIES TO WATCH A MOVIE – YOU CAN SEE THE RECENT RELEASES AT HOME, EVEN ON YOUR SMARTPHONE.

But there are nice things about going out to the movies including the shared experience.

And speaking of shared experience, it seems that we collectively emit chemicals when something powerful happens on the screen – and these chemicals can be measured by analysing the air coming off the audience.

AIRBORNE CHEMICAL COMMUNICATION

It kind of makes sense. Many creatures use chemicals floating through the air to communicate – plant to plant, plant to insect, insect to insect.

It's tricky to verify this effect in humans. On one hand, there are all kinds of claims for human airborne chemical communication – armpit and sweat odours affecting menstrual cycles and feelings of fear, tears affecting testosterone levels, and sleeping babies responding to the smell of a lactating breast. On the other hand, there are very few large-scale, reliable studies backing up these claims.

TO THE MOVIES

Mind you, you need some pretty sophisticated equipment to measure subtle changes in chemicals in the air.

The team of scientists who did this study at the movies, usually spent their time flying over pristine Amazon forests, sampling the volatile chemicals given off. So back in Mainz, Germany, on Christmas holidays, the obvious thing to do was to keep working! The scientists wandered down to the local cinema, plugged in their fancy chemical analysis equipment, and measured the air coming out at the theatre. The incoming air was pumped in under the audience's seats, and exited though vents in the ceiling.

Over Christmas 2013, the team analysed the air from over 9500 cinemagoers who watched 108 screenings of 16 different films. Over 870 volatile chemicals have been identified in human breath, but they concentrated on just 100 of them.

The film genres that produced the most recognisable and consistent chemical signatures were suspense and comedy. From the point of view of Evolutionary Biology, "suspense" corresponds to "be alert", while "comedy" links to "stand down".

One chemical that stood out in the early analysis was isoprene (C_5H_8). This chemical is stored in your muscles and is released when your muscles contract. It's also involved with the production of cortisol – which is part of the Fight-or-Flight Response. (And just to show how complicated the human body is, it's also involved in the synthesis of cholesterol.)

CHEMICAL SIGNATURES

On each of four successive days, there was a peak in the isoprene coming out of the movie theatre air conditioning ducts when the audience stood up to leave the theatre – as you would expect. After all, they were using their muscles to stand up and walk out.

But the surprise was what happened *during* one of the movies, *The Hunger Games 2*. At exactly 2.58 p.m. on each of the four successive days, there was a clearly discernible jump in the isoprene – corresponding precisely to when the heroine's dress caught fire. And then there was another sudden jump at 3.07 p.m., when the final dramatic battle began. It appears that when the dress caught on fire and when the big battle began, the audience, without realising it, tensed their muscles in unison.

At both of these events in the movie action, and exactly in synchrony with the jump in isoprene, there was a jump in carbon dioxide (CO_2). Perhaps the audience, in addition to tensing up, were breathing faster.

There are so many fields where this new technology could be applied – psychology and biology, the making and marketing of films, and of course, in the diagnosis of disease.

I wonder if *The Hunger Games 2* was as good for the scientists stuck outside the movie theatre as it was for the audience inside? May the odours be ever in your favour...

Very unlikely, unless the glass has been pre-scratched (e.g., diamond ring). We tried in lab, big amplifier, no luck.

The Haemophile Cafe

Today's Menu

Type O Lumberjack — $21
Type A pregnant lady — $29
Type B orphan — $19
Type AB patient — $15

24

MOZZIES LOVE (SOME) HUMANS

WHY DO SOME PEOPLE COMPLAIN THAT THEY'RE MOZZIE MAGNETS, WHILE OTHERS REMAIN BLISSFULLY BITE-FREE?

Answer Number One: it's chemistry!

Answer Number Two: anyone can get bitten, but not everybody reacts to the bite.

MOZZIE 101

Only the female mosquito bites you. No hard feelings, it's just business – she needs the protein in your blood for her babies.

Mosquito bites alone would be mostly just an annoyance. But the mosquito herself can be infected – and pass that infection on to you.

Different mosquitoes carry different germs.

The mosquito *Aedes aegypti* can carry dengue and the Zika virus. A different species of mosquito, *Anopheles gambiae*, can carry malaria. In 2015, malaria infected 214 million people and killed 306,000 of them, the vast majority being children younger than five. Another mosquito, *Aedes vigilax*, carries the Ross River Virus.

LOVE AND HATE

Mostly, mosquitoes love heat and carbon dioxide. Once mozzies get close enough to you, they can be either attracted or repelled by any of the several hundred chemicals you exude. These chemicals can come from your diet, from bacteria on your skin, and from your genetic inheritance.

It's early days still for learning about those several hundred chemicals you emit.

One study looked at identical twins (whose DNA is virtually the same) and non-identical twins (whose DNA is partly the same and partly different). Mosquitoes were equally attracted to each identical twin in a pair, but this didn't happen with the non-identical twins. So your genes affect how much mosquitoes love you.

Eating garlic or vitamin B_1 does not repel mosquitoes. We know that the *Aedes aegypti* mosquito loves the smell of lactic acid – so plan not to exercise in her wetlands. At least one species of mosquito loves blood group O – but we don't know how all the other species feel about blood group O (or, indeed, any other blood group).

Mosquitoes love people who drink beer, and they prefer pregnant women to non-pregnant women because their core body temperature is higher. We know that men are more attractive to mozzies – because their greater body mass is linked to more heat and more carbon dioxide emission.

Darker colours can absorb heat better than lighter colours, making you more attractive to mozzies. We also know that hairy arms or legs can be a physical deterrent to mozzies.

Mozzies can be fussy. So *Anopheles gambiae* is attracted to Limburger cheese. By a coincidence, the bacteria that make this cheesy smell are closely related to the bacteria on some people's feet.

People who are already infected with malaria become even more attractive to *Anopheles gambiae* at a very specific stage. This is when the malaria parasite is more transmissible to another person.

Mosquito saliva is complicated. Some people react very strongly to certain chemicals in the saliva – and so complain bitterly about being bitten. But not reacting is different from not being bitten. They could still get bitten, but would not know it. They might think that they are unattractive to mosquitoes – but that is incorrect. In that case, they might not apply DEET (the best repellent/high-powered insecticide), get bitten without knowing it, and then suffer a mosquito-borne infection (dengue fever, malaria, Zika virus, etc.).

DEET works to repel mosquitoes in two ways – by its odour in the air, and by direct contact when the mozzie tries to land on your skin. By the way, the protection from DEET lasts for hours, while that from citronella, a common alternative repellent, lasts only minutes.

So remember, mozzies might make you sick – even if they don't make a mark.

Potential light aircraft hazard → paint white. FAA regulations: "As long as the wind turbines are painted white in colour, daytime illumination is not required."

Double or Nothing

25

DOUBLE-YOLK EGGS

WHAT IS THE CHANCE OF GETTING A SINGLE DOUBLE-YOLK EGG?

Turns out to be about one in a thousand.

And the chances of four double-yolk eggs in a row? Under the right circumstances, they can be surprisingly high.

LAYING THE EGG

Five regions of the oviduct – the tube from the ovary to outside – are involved when a chicken lays an egg.

The first region (the infundibulum) is where the egg yolk arrives, after leaving the ovary. In region two (the magnum) the yolk is coated with about half of its white albumen. In region three (the isthmus), more albumen is added – and the inner and outer shell membranes are deposited. The eggshell is added in region four (the uterus). The egg leaves through the fifth region, the vagina. The uterus virtually turns itself inside out to deliver a sparkling clean egg, free of faeces – even though chickens use the passage for both eggs and poo.

DOUBLE YOLK

You get a double yolk when the ovary is too "keen". It produces one yolk, and instead of waiting one day as normal, it quickly produces another yolk. The two yolks then move together through the rest of the process, and get laid as a single twin-yolk egg.

Double yolks happen more frequently in young chickens, when they're in their early days of egg production – 20 to 28 weeks old. These young chickens, still getting the hang of being a grown-up hen, produce a double-yolk egg once every 100 lays. When you average out the number of double-yolk eggs from chickens of all ages, about one out of every 1000 eggs has two yolks. So, ignoring all other factors, the chances of getting four double-yolk eggs in a row from a single carton should be: $1/1000 \times 1/1000 \times 1/1000 \times 1/1000$ – or one in every trillion.

WHY DOUBLE YOLK?

But there are other factors – lots of them.

First, double-yolk eggs are usually larger than single-yolk eggs. Eggs are usually sold in the sizes of small, medium and large. So if you find a double-yolk egg in a carton of large eggs, the chances are higher than normal that another double-yolk egg could be found.

Second, the eggs in any given cardboard carton are likely to have come from the same flock. Chickens in the same flock are usually the same age. Again, this increases your chances of getting a second double-yolk egg, if you've already picked the first one.

GUARANTEED DOUBLE YOLK

In the UK, department store Marks & Spencer have increased the odds to 100 per cent. In 2015, they launched boxes of six extra-large eggs – each one guaranteed to have a double yolk.

They started with young chickens, and then "candled" their extra-large eggs. "Candling" is a technique usually done in a darkened room. A bright light is shone through the big end of the egg. Candling can detect blood in the whites, cracked shells – and double yolks. The cost is 2–3 times higher – you have to shell out £2.75 for the six eggs, or about AU$5.

TWO YOLKS = TWO CHICKENS?

Unfortunately, they rarely hatch.

In a sat-on-and-hatched-by-Mum egg, the baby chicken has to rotate around, so it gets its head up to where the air cell is (the round end). Then, under normal circumstances, the baby chicken will peck its way out.

But if there are two chickens inside, they will almost invariably fight each other. Neither of them will be able to get to the air cell, so they'll both die.

However, there have been a few very rare cases where the egg has been very carefully opened at exactly the right time (in a kind of mini caesarean), and two chickens have survived from a double-yolk egg.

It makes me wonder: if hens can give us double yolks, why can't Easter Bunnies give us double-chocolate eggs?

No. Rainbow = light sent to retina by raindrops @ 42 degrees away from your shadow. Lightning striking a rainbow is not real but "virtual", i.e. coincidence.

26 GOD, CAFFEINE & CHOCOLATE

COFFEE AND CHOCOLATE ARE PART OF LIFE'S SUPREME PLEASURES. YOUR HEART AND SOUL AND TUMMY ALL THINK SO – AND NOW BIOCHEMISTRY PROVES IT.

And just for fun, coffee and chocolate have proposed a solution to one of the philosophical mysteries that has haunted deep thinkers for millennia.

The active ingredient in coffee is a chemical that is called caffeine by the general population. However, the Biochemists call it 1,3,7-trimethylxanthine. If you look at the molecule of caffeine you see what's called the xanthine section in the middle – and you can see three methyl (CH_3) groups at the 1, 3, and 7 positions.

Caffeine is removed from the body fairly quickly – about half is gone after about three to four hours. We call it a vasoconstrictor because it tends to close down your blood vessels.

Caffeine is remarkably similar to the active ingredient in chocolate, theobromine. Chocolate is a heavenly food, and the name reflects this. In "theobromine" the "theo" means "God" (as in "theology") while "broma" means "food" – literally Food of the Gods. (Despite the name, there is no bromine in "theobromine".) In general, the darker the chocolate, the more theobromine it contains. Theobromine hangs around for roughly twice as long as caffeine – its half-life is about seven hours.

Caffeine chemical structure

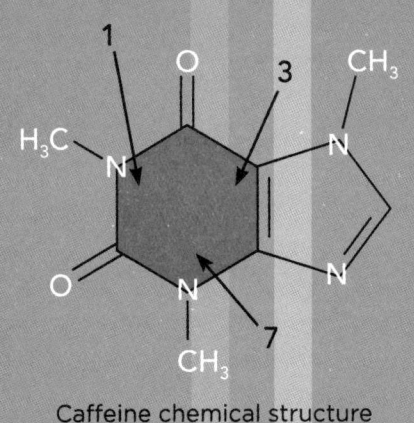

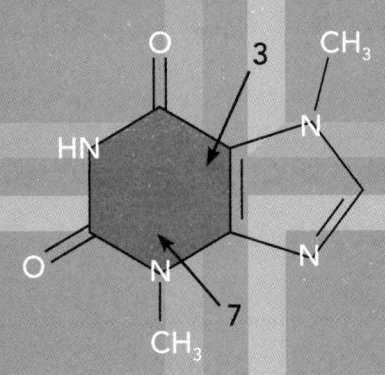

Chocolate chemical structure

The chemical formula of theobromine is 3,7-dimethylxanthine. It's almost identical to caffeine. There is just a tiny difference. Theobromine, as compared to caffeine, is missing a methyl group at the 1 position. And wait for it – theobromine opens up blood vessels. It's a vasodilator.

Two things are obvious.

One, whenever you drink coffee, you should also have a little chocolate to balance things out.

Two, this is no coincidence. It solves the Problem of the Ages. Yes, if God exists, then God wants us to have chocolate with coffee.

1) Universe is old, you are young and short-lived 2) Expansion is overcome by Gravity 3) Expansion is between galaxies.

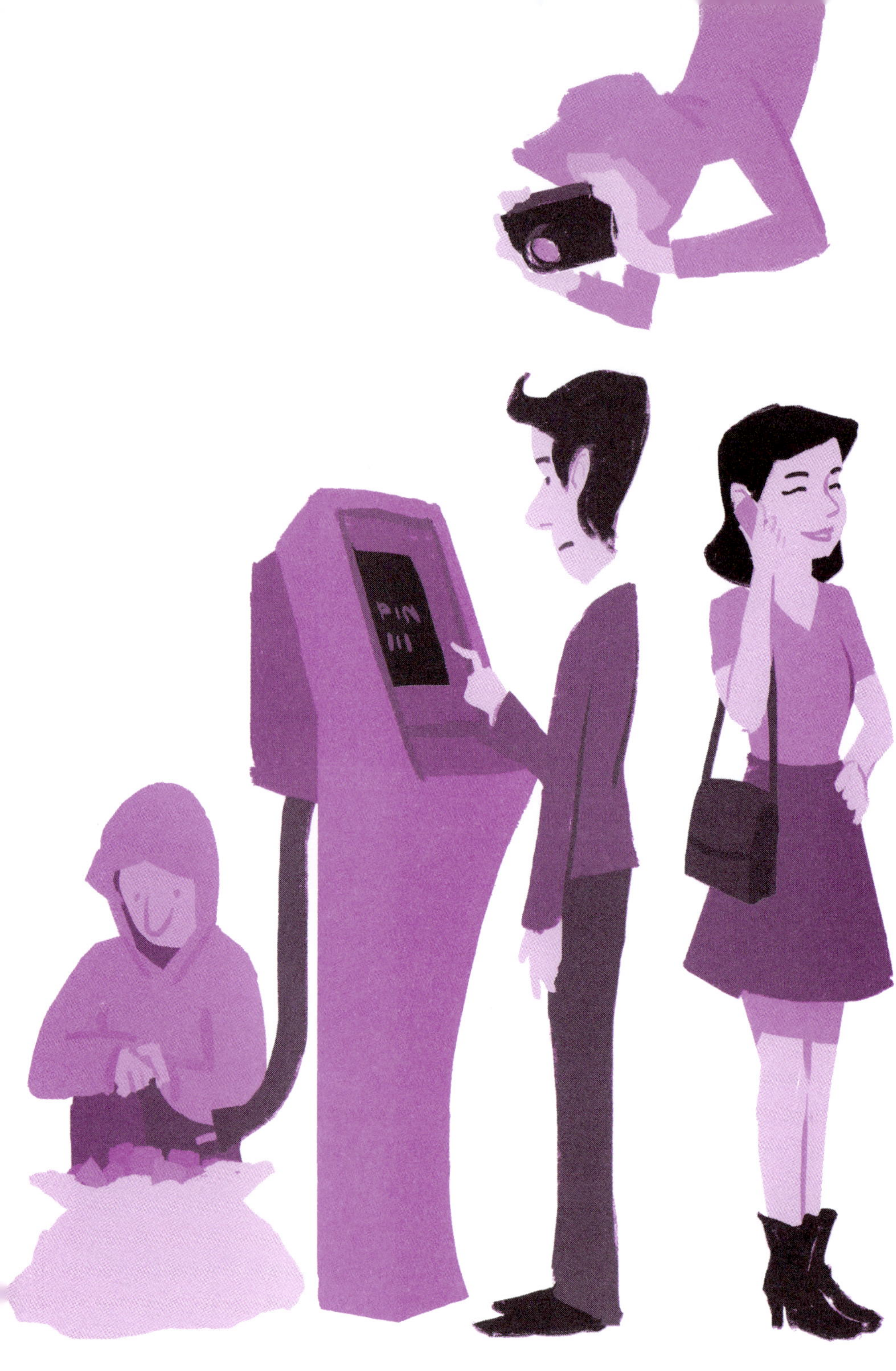
PIN

27

CREDIT CARD THEFT

ALONG WITH A MILLION OR SO OTHER AUSTRALIANS, I HAVE BEEN THE VICTIM OF CREDIT CARD FRAUD.

In theory, I could have gone to the Dark Web and bought back my stolen credit card information – from people certified to be genuinely crooked (no honest people allowed). To get onto the Dark Web, I would have used software that was developed by the US Naval Research Centre – yes, taxpayer dollars at work!

Welcome to the bizarre Land of Credit Card Theft, where something can be taken from you, even though you still have it. That "something" is your credit card information.

MAGNETIC STRIPE AND CLONING

The magnetic stripe on the back of your credit card is based on ancient technology – over half a century old. It has major security and fraud issues.

For example, using simple technology that you can buy for only $100, you can replicate the information from your card's stripe onto another card with a blank magnetic stripe. (The MSR606 is very popular – google it.) This card would be flat (no raised

MY STOLEN CREDIT CARDS

On two separate occasions, my bank has rung me, asking me if I had just made some purchases in the town of Hendersonville, just outside Las Vegas in Nevada in the USA. I said "No". (I wonder why Hendersonville. Has the same person bought my stolen credit card details twice?)

In each case, the person using my stolen card had made a purchase of just a few dollars, to see if my credit card actually worked. Then they immediately made a series of purchases, each just under $1000. And then my bank rang me and "killed" that credit card – and that spending spree.

HALF A MILLION HACKS PER DAY

In an interview with Sienna Kossman, Greg Wooten, the CEO of the fraud prevention technology company SecureBuy said, "about half a million data resources [are] being breached each day. The hackers extract the data, and analyse it using analytics to match up information the best they can, and then monetise for the highest value possible when they go to wholesale it."

@DoctorKarl Could 5 EMP devices destroy all the electronics on the planet?

writing) and devoid of all other writing or images. It wouldn't look like a real credit card. In fact, it would look just like the blank cards that hotels give you to open the door to your room. However, you could use this blank at an ATM, or in a shop where you swipe the card yourself – where there is no other human involved who might notice that your card doesn't look right.

But if you want to make a card that you can take into a shop and hand to the staff, you have to spend up to $1000. This will get you a Card Embossing Machine. This counterfeit card is not flat and blank – it has raised writing for your name and the numbers, the appropriate colours, and even a hologram of the issuing bank.

CREDIT CARD THEFT 101

The Australian Criminal Intelligence Commission defines Credit Card Fraud as the "fraudulent acquisition, and/or the use of a card, or the card details, for financial gain". But it affects more than credit cards. This criminal fraud can include Debit Cards, Charge Cards, Gift Cards, and ATM Cards. In the Australian Financial Year of 2012–2013, about 1.4 million acts of fraud were carried out using Australian-issued cards. The value of the transactions was around $280 million.

The hidden information on your credit card can be stored in three main ways.

First, there's the old-fashioned magnetic stripe. The security here is the store employee checking your signature on the transaction receipt matches the one on the back of the card, to verify the card is yours.

Second, there's the chip. It needs your Personal Identification Number (PIN) for authentication.

More recently, there's the Tap-and-Go or PayWave card, which carries an internal Radio Frequency Identification (RFID) tag. You don't need to authenticate this card for smaller transactions (typically under $100). But you do have to enter your PIN for larger transactions.

EMP device = "small" thermonuclear bomb in low Earth orbit → gamma rays → destroy most electronics. Need about 2–5/continent.

STEALING DUMPS

The thieves call the stolen credit card information a "dump". How do criminals steal this credit card information? Many ways.

First, the hackers can compromise a high-value network. This could belong to a company that processes transactions between the various banks involved in a transaction – such as the bank that issued your card, and the different bank used by your retailer. If hackers can get into the network, they can steal an enormous number of card details in a very short time.

Second, they can get into the database or website of an online merchant. The merchant might have stored the credit card details of all their customers.

The third method is an oldie, but a goldie – "skimming". An ATM has a slot for you to insert your card. The crooks can attach a physical device (the skimmer) onto the top of this slot – sometimes just with double-sided tape. The skimmer has its own matching slot, some electronics and a tiny camera. So now your credit card goes through two slots – first the fraudulent one, and then the real one so that you can get your cash, or pay for your petrol. The electronics in the skimmer capture all your card's details, and the camera films you keying in your PIN. Sometimes, the electronics send the stolen information via WiFi or Bluetooth to a crook nearby. In some cases, the software is fine-tuned to ignore all except gold and platinum cards, because these tend to have a higher limit.

A variation on "fixed" skimming is when a waiter or taxi-driver might carry a small skimmer the size of a matchbox or an ice-cube. They take your card, smoothly and unobtrusively swipe it through their portable skimmer, and then swipe your card in the official Point-Of-Sale Device.

Other hacking methods include malware that has been installed on your computer or on a public computer, crooked employees, and even the old-fashioned theft of the physical card.

@DoctorKarl Does the Moon orbit the Earth in an ellipse?

But it's the hacking of computer systems that delivers the biggest and quickest returns to the crims. So in 2013 alone, Adobe Systems had 152 million data records stolen, Experian (a company that does credit history and offers free credit reports) had 200 million data records stolen, while that friendly retail giant, Target, had 40 million credit card numbers and various identification details stolen.

But how do the criminals, having stolen millions of credit card details, turn these dumps into dosh? They need different skills – in fact, they need Organised Crime.

DUMPS TO DOSH? ORGANISATION

The processes needed to turn "numbers" into "bucks" are complex, and rather specialised. Just like the criminal underground of the Mafia back in the 1970s, we're again dealing with organised crime.

Furthermore, there's a sense of urgency. Stolen credit card details are a bit like fruit – they are highly perishable. You need to convert them into cash as soon as possible. For example, in the case of the 40 million card details stolen from Target back in 2013, the details lost 70 per cent of their "value" in just two months.

How much value do dumps have? They're sold for between US$1 and $100 – depending on the type of card, its credit limit, how much personal information is supplied (phone number, date of birth, etc.), how many of the cards in a batch offered for sale are still valid, and so on. Two surprisingly useful pieces of information are where the real owners of the credit cards live, and their buying patterns. This means that the final user can make purchases that look in-character to the cardholder's purchases, and don't immediately attract attention.

(HIGHLY) ORGANISED CRIME

Cybercriminals have set up highly developed and well-funded organisations. They are very similar to legitimate companies. There are many people involved, with different kinds of expertise.

1) All orbits elliptical 2) Moon doesn't orbit Earth. They orbit their common Centre of Gravity ~1410 km under equator.

Some people write the malware that allows hacking into supposedly secure data banks.

Others will do the actual hacking. For example, they might bust their way into a big merchant's Point-Of-Sale network and steal the data while it's still being processed – before it gets sent to the bank.

Following them will be a separate team who collect and check the validity of the stolen data. After all, the bank might have already cancelled the card.

Another team posts the cards on so-called "Dark" websites. That's where you find the sellers, and the buyers, of the dumps (stolen card details).

After them come the card counterfeiters. This is for those situations you want to use a physical card, as opposed to using your dump on the net. (Physical cards were used with my stolen details in Hendersonville, Nevada.)

Finally, we arrive at the end user of the stolen card.

Just like a legitimate business, there are several layers of management, and, apparently, a ready supply of workers. These organisations have access to skills and technology resources that rival those of large legit companies.

UPSIDE-DOWN TOWN

You can buy dumps in bulk, on the Dark Web. It's a crazy upside-down, back-to-front world.

If you want to do business on the regular web, you need to prove that you are honest. But before you can go to a site that sells stolen credit cards, you have to prove the opposite – that you are truly dishonest.

A typical site might have a log-in screen, with a fan of golden credit cards superimposed onto a black background. You have to sign in with a username and password. But how do you get these? You have to convince two people who have previously bought dumps to email the website moderator to vouch for your criminality – to say that (honestly) you are a complete crook.

THE DARK WEB AND TOR

The Dark Web became (in)famous because of the Silk Road – a website where users could buy drugs, and other illegal products and services. The Dark Web (also called the "Dark Net") is intimately linked to Tor.

Tor was developed in the mid-1990s by mathematicians and scientists at the United States Naval Research Laboratory. It was finally launched on 20 September 2002. Its purpose was to allow US Intelligence Officers to communicate anonymously online.

Tor stands for "The Onion Router". It allows an internet user to conceal their location and internet usage. The Tor Browser is about 40 megabytes in size, is based on Firefox, and works on all major computer platforms (Mac, Windows, Unix, etc.).

Normally, you send and receive many packets of information to and from the Internet. Each packet usually carries a small amount of information, including the address of where it came from, and where it should go.

But in Tor, all this data is encrypted. Furthermore, there are many "layers" of information, like the layers of an onion.

So your packet is deliberately sent to a randomly selected Tor Relay Station (there are about 7000 of them). Here, the outer layer is stripped off and decrypted. Following the instructions in this layer, the packet is sent to another Tor Relay Station. The process is repeated many, many times. The disadvantage is that this slows down your web browsing. The advantage is that it makes your web activities anonymous.

Tor is used by military professionals, buyers and sellers of illegal products, legal and financial services, whistle-blowers, civilians who like their privacy, users of political chat rooms and information databases for journalists, and Facebook users in Nauru.

In April 2015, Nauru banned Facebook. So Facebook set up a Dark Web website that could be accessed by Tor. It's facebookcorewwwi.onion.

Ice. Jet fuel (atoms of C, H, & O in "kerosene") burns → CO_2 (invisible) + H_2O (usually invisible, but can → ice crystals).

It's like getting two references when you apply for a job – but backwards. Here, you need two references saying how bad you are. These documents, in the wrong (or is it right?) hands, could land you in jail.

Once you get into a dump site, it's upside-down town again – the background is black, instead of white (nice shady touch). Alongside the main business of buying stolen credit cards, there are banner advertisements for all kinds of illegal activities – such as software for hacking and phishing, and of course, tutorials in how to use this illegal malware.

Just as with the big legit traders such as eBay and Amazon, this illegal site is rated by its customers. They might write that they bought 50 or 5000 cards, and that they were a good mix of basic, gold and platinum cards. So one set of criminals (the card customers) is rating the honesty of another set of criminals (the card sellers).

There's even a section for Frequently Asked Questions (FAQs). "Do you ship in bulk? Yes. If I become a repeat customer, will I get a discount? Yes." That sort of thing...

Then, you have to click on a box that you agree to the Terms and Conditions. You have to certify that you really are a criminal – and that you are specifically neither a journalist, nor a law enforcement officer.

One site selling dumps had a rule that if you used CAPS LOCK (what they called "shouting" on email), you would be suspended for a week.

So it is OK to buy and sell millions of dollars of dumps (as well as drugs, flamethrowers, and possibly small tactical nuclear devices). But it's not OK to use ALL CAPS.

BASIC ECONOMICS – JUST BUSINESS

Once you click OK, you're inside the dump site.

And there they are – thousands of stolen credit card numbers, with their three-digit security codes. There are thousands, rather than tens of thousands, because the sellers don't want to release too many dumps at once. The crims know the Law of Supply and Demand, and don't want to drive the price down too far.

How do you pay for dumps? Well, definitely *not* with a credit card – after all, it might be stolen. (Oh, the irony – who can you trust nowadays?) Instead, you use a digital currency, such as bitcoin (see story on page 129) or WebMoney.

But suppose that you are a victim. If you have to notify your bank that fraudulent purchases have been made on your credit card, you normally don't have to pay for those items or services. The bank, or the merchant, will carry the cost.

Sounds like the perfect victimless crime? No.

You see, ultimately, the bank and the merchant pass the cost back to you, the consumer, by way of higher bank fees and higher service prices. But, no hard feelings, that's just business . . .

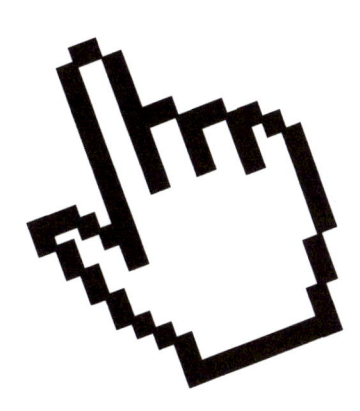

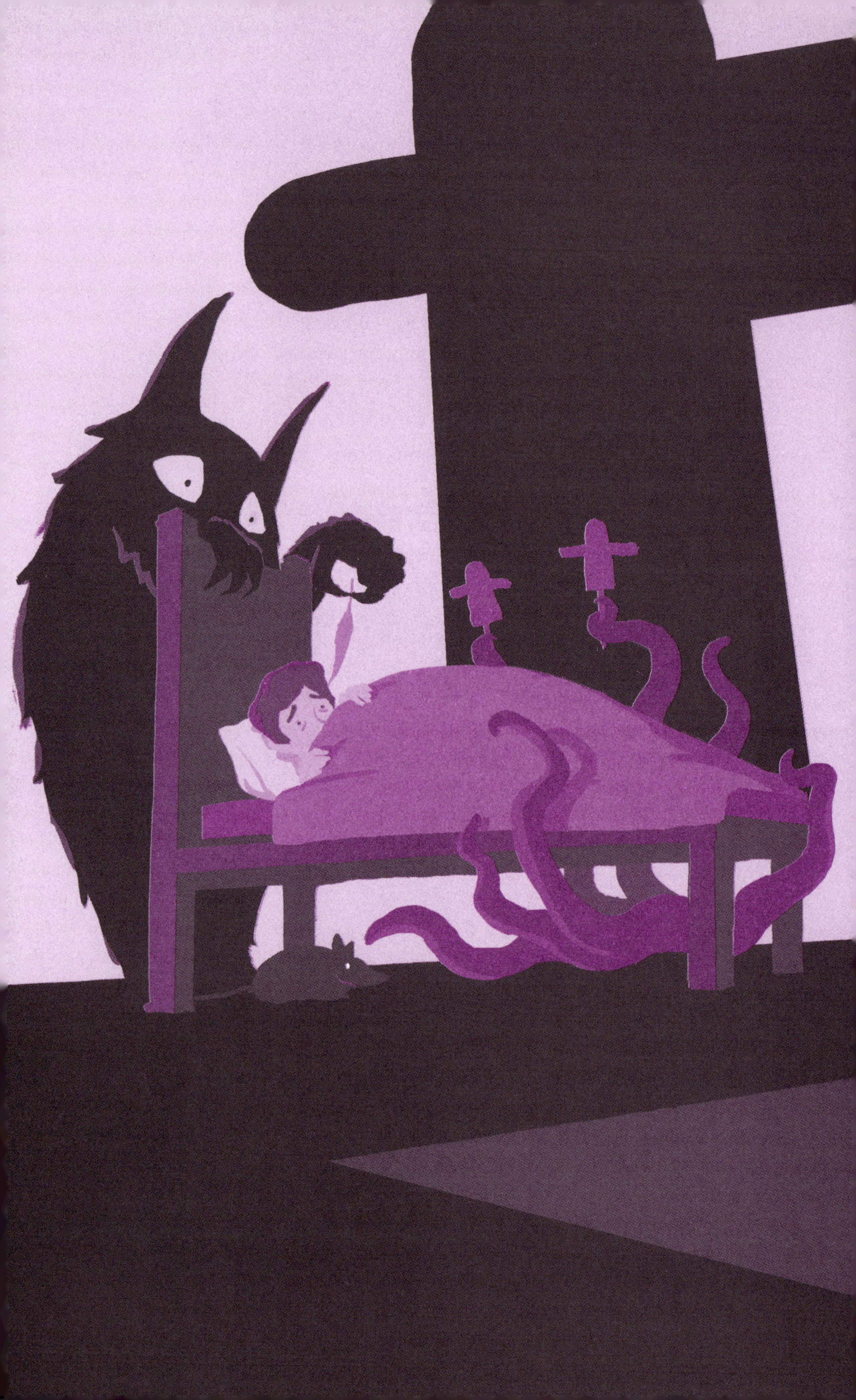

28
SLEEP BADLY IN AN UNFAMILIAR BED

YOU'VE GONE ON HOLIDAY TO AN ABSOLUTELY FABULOUS DESTINATION WHERE EVERYTHING IS JUST PERFECT. WHY IS YOUR SLEEP ROCKY ON THE FIRST NIGHT?

Recent research tells us it might be caused by the left side of your brain staying partly awake, vigilantly alert for threats. On that first night, it's a light sleeper.

FIRST-NIGHT EFFECT

Welcome to the First-Night Effect, well known to Sleep Scientists. Sleep Scientists usually discard the data for the first night's sleep of volunteers in their sleep laboratory, precisely because their sleep is so restless and disturbed. "Why does this happen?" and "What is happening inside the brain?" were the questions Yuka Sasaki of Brown University in Rhode Island asked herself.

She and her Sleep Scientist colleagues recruited 35 healthy young volunteers. They monitored their brain activity on two nights in her sleep laboratory.

Now here's a clue as to what they found. Dolphins and some birds are known to sleep with half a brain at a time, while the other hemisphere stays awake. This lets them simultaneously both snooze and migrate-and-navigate. But we had never seen this in humans before.

NEUROLOGICAL NIGHT WATCH

On the first night, their volunteers didn't sleep particularly well, and took longer to fall asleep. In addition, their left hemispheres stayed more "awake" and kept watch on the outside world. This is quite unusual – normally, in humans, both hemispheres fall asleep in synchrony with each other.

There are many dedicated networks in the human brain. One of them is called the Default Mode Network (DMN). This collection of nerve cells become active when you are awake and just hanging around – like when you daydream and have creative thoughts. The DMN also becomes active when you sleep.

On the first night of sleep in the lab, the DMN in the left side of the brain showed less of its typical sleep-related activity. In other words, on that first night in an unfamiliar bed, your left brain kind of stays more awake.

There was another finding that showed that the left brain was more alert. The Sleep Scientists played sounds into the right ears of their volunteers. Because of the crossover that happens in the human brain, sounds played to the right ear are processed by the DMN in the left brain. In the sleep lab on that first night, when sounds were played to the right ear, the volunteers were more likely to be roused, and to wake up more quickly, than when the sounds were played to the left ear.

The disadvantage of being more rousable on that first night is that you won't get a perfect sleep. But the advantage is that you are more likely to survive a threat. This pathway might be related to parents being able to sleep through a thunderstorm, but snapping awake as soon as their newborn baby makes the slightest noise.

COMPLAINTS

One problem with this study was that the scientists monitored only the first sleep cycle of the night, not all four to six of them. Another was that the study didn't see what happened with only left-handed volunteers – who have slightly different left–right brain wiring.

First studies are never perfect. That's why you have follow-up studies.

SOLUTION

How can you minimise the First-Night Effect and get a good night's rest? You could take your favourite pillow with you when travelling. Keep to a normal rhythm and avoid the mini-bar to help you sleep better.

And if you belong to a Hotel Loyalty Program, not only will you get lots of points, the rooms within the same hotel chain will be slightly familiar.

Incorrect. "Gravity" is created by "Mass". (And the Higgs Field and Higgs Boson are involved, Nobel Prize 2013.)

29 SMOOTHIE SCAM

"Smoothie" is a lovely soothing name – so from a marketing point of view, this drink is off to a good start. It's a thick beverage, blended from raw vegetables or fruit, with added ingredients such as dairy products, water, ice and sweeteners.

How about this for a question: do you absorb more energy (calories or kilojoules) when you drink a smoothie, as compared to eating the same ingredients?

Almost certainly yes, according to Sarah B. Krieger, a Dietician and spokesperson for the US Academy of Nutrition and Dietetics.

Fibre in whole fruit and vegetables "acts as a net" to slow down the process where the natural sugars in food turns into sugar in your blood. Yes, the fibre is still there in your smoothie. But when the fruit and veg were blended by sharp high-speed blades, the fibre was also pulverised. It's not as effective as non-pulverised fibre at slowing down digestion. So you are more likely to feel hungry, sooner – as compared to eating the same quantity of fruit and vegetables.

And let's not forget quantity. You can easily drink in a minute or so what it would take you 15 or 20 minutes to chew. It's very easy to drink the chilled juice of five oranges on a hot day – it's a lot harder to eat them. So you can be taking in a whole lot more energy by drinking, than if you had simply eaten some fruit and vegetables until you were full.

@DoctorKarl Are earthquakes in Japan linked to earthquakes in Ecuador?

Another pitfall is the smoothie you don't make for yourself: commercially prepared or store-bought smoothies. These often contain added sweeteners and high-energy ingredients to make them more filling. So a Smoothie King 20-ounce (590-millilitre) Hulk Strawberry contains butter pecan ice cream – and delivers nearly 4200 kilojoules, about half your daily requirements. It isn't even prepared from whole fruit – it has reconstituted strawberry puree, white grape juice concentrate, protein powder and more.

No. 1) too far apart 2) different types of earthquake 3) still very low numbers of earthquakes/year.

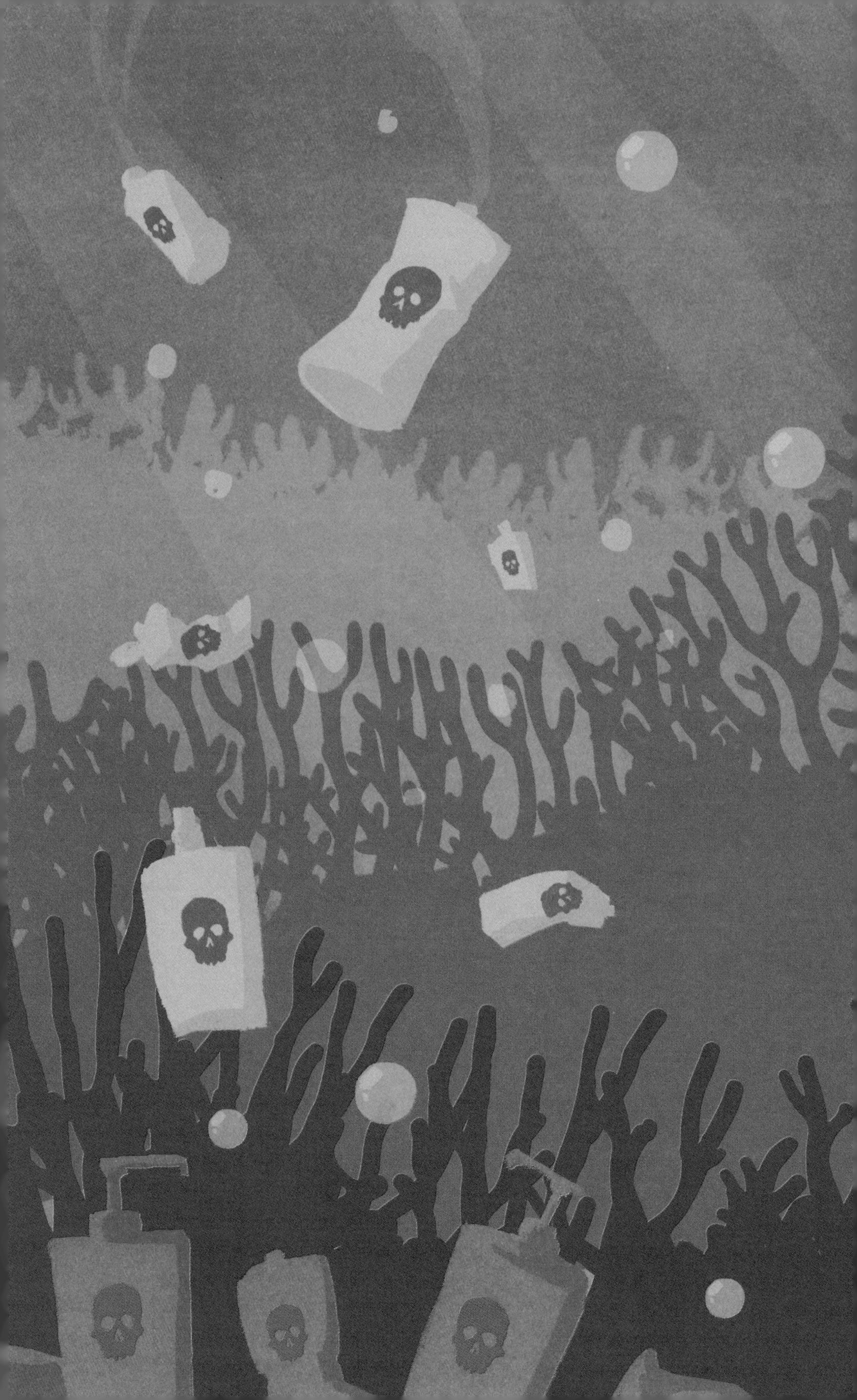

30
SUNSCREEN ATTACKS CORAL REEF

FROM CRADLE TO GRAVE, AUSTRALIANS USE SUNSCREEN TO AVOID SUNBURN AND SKIN CANCERS. BUT EVERYTHING HAS A COST.

Unfortunately, one popular sunscreen chemical seems to attack baby coral.

GALAPAGOS ISLANDS

In 2015, my family and I went to the Galapagos Islands. Each day, our boat took us to a different location.

I was very surprised when our guide asked us to not wear sunscreen at one particular coral reef. He said it was a very fragile reef, and quite small, and that the local currents tended to trap contaminants.

OXYBENZONE 101

The chemical is oxybenzone – one of the most widely used organic UVA filters. It's also called benzophenone-3 (BP-3). It absorbs UVB, so it blocks UV radiation ranging from 270 to 350 nanometres. (Not many organic sunscreens block both UVA and UVB.) Oxybenzone is used in hairsprays, cosmetics, nail polishes – and in over 3500 sunscreen preparations worldwide.

Oxybenzone is very fat-loving (lipophilic), so it penetrates the skin easily. It acts on humans to be an "endocrine disruptor". This sounds bad – endocrine chemicals, better known as hormones, are essential for the body to work properly. However, the effects (so far) seem to be (probably) minor.

On average, oxybenzone will appear in the urine of over 96 per cent of people who apply it to their skin. Most developed countries allow it to be used at levels up to 5 to 10 per cent of a sunscreen's ingredients. However, the Swedish Research Council deems oxybenzone to be unsuitable for children under the age of two, who have not yet developed the enzymes to break it down.

This was the first time that I had ever heard that sunscreen could harm a coral reef.

OXYBENZONE ATTACKS CORAL

Oxybenzone has four separate bad effects on coral.

First, it predisposes coral to bleaching at lower temperatures than normal. If the ocean is hotter than normal, then there's even more coral bleaching. (See my story "Great Barrier Reef" in my 38th book, *Dr Karl's Short Back & Science*.)

Second, oxybenzone damages the DNA of coral in many different ways. Doctor C. Downs and his team say this includes "oxidative damage to the DNA, formation of cyclobutane pyrimidinic dimers, single-strand DNA breaks, cross-linking of DNA to proteins", and so on and so on. The overall result is that coral with damaged DNA are less able to reproduce. If they can reproduce, their offspring are likely to be unhealthy. This leads to each generation of coral being less healthy than the one before.

Third, oxybenzone acts as a powerful endocrine disruptor. The juvenile coral produce too much calcium carbonate and end up totally encasing themselves inside their own skeleton. This leads to their death.

Finally, oxybenzone deforms the juvenile coral. They stop swimming, change shape, and their mouths grow five times larger.

Sports drinks can replace salts IF you have generated LOTS of sweat, for a LONG time, i.e. top athlete. For the rest of us, no advantage.

OXYBENZONE AND CORAL

Each year, about 6000 to 14,000 tonnes of sunscreen enter the oceans. After all, you should slather on generous amounts of sunscreen.

Oxybenzone can be toxic to baby coral at levels as low as 62 parts per trillion. In plain English, that's equivalent to one drop in 6.5 Olympic swimming pools. Doctor C. Downs and his team surveyed reefs in Hawaii and the US Virgin Islands. They measured oxybenzone levels as high as 1,400,000 parts per trillion.

SOLUTION

What can we do? As an example, consider chlorofluorocarbon gases, also known as CFCs.

When we realised that CFCs were damaging the Ozone Layer, the chemical companies initially said that effective substitutes were impossible. However, once the governments of the world put pressure on them (via the Montreal Protocol), effective substitutes were soon found.

The solution is straightforward. People still need to avoid skin damage (up to and including skin cancer). Sunscreen is part of that strategy. We know that certain ingredients used in sunscreens, such as titanium oxide and zinc oxide, are not as harmful to coral as oxybenzone and there are probably other chemicals we haven't discovered yet. So we could find chemicals that don't harm humans and ocean creatures – or we could cover up with high-SPF clothing, and avoid using sunscreen altogether.

@DoctorKarl **How can I test mercury levels from amalgam tooth fillings?**

Sunscreen Attacks Coral Reef **< 187**

**See GP for blood test, possible referral to toxicologist.
Overwhelmingly, Hg leakage from amalgam is non-toxic.**

31 COFFEE'S A DIURETIC?

Worldwide, the daily consumption of coffee is enormous – 1.6 billion cups. Tea is even more popular, with double that consumption. But while both tea and coffee can be refreshing and invigorating, there is a niggling worry. Can they be counted in your daily liquid consumption – or can't they, because they are supposedly dehydrating? (And it's a myth that you shrivel up into dust if you don't drink at least Eight Glasses of Water Per Day – see my 34th book, *Game of Knowns*.)

The scientific evidence on the diuretic effect of tea and coffee is not entirely clear. It seems that if you're not a regular tea or coffee drinker, they might be mildly diuretic. But after a few days, they don't keep on working that way.

COFFEE SCIENCE BEGINS IN 1928

One of the earliest studies was carried out in 1928 with an incredibly small sample size – just three men. Over two successive winters, they had their urine volume measured. Sometimes they drank water, sometimes coffee, sometimes tea, and sometimes they drank water that had pure caffeine added. When, after a two-month abstinence from tea or coffee, they drank caffeinated water, the volume of their urine increased by 50 per cent. However, very shortly after resuming regular caffeine consumption, the volume of urine returned to what was normal for them.

A 1997 study on just 12 volunteers kept them off caffeine for six days. On the seventh day, they drank six cups of coffee – which increased the

@DoctorKarl What makes people bruise more than others?

urine volume by 750 millilitres per day. But the study didn't follow them further to see if their urine production dropped back to normal over time.

One study in 2000 looked at just 18 healthy adult males (aged 24 to 39). They found "no significant differences in the effect of various combinations of beverages on hydration status of healthy adult males".

A 2002 meta-study of 10 other studies found that if caffeine is a diuretic, it's a very mild one at most. About 80 per cent of comparisons within the various studies showed that the presence of caffeine in the water (carried in either tea or coffee) had no effect on the volume of urine produced.

A UK study in 2011 revealed "no significant differences between tea and water for any of the mean blood or urine measurements. It was concluded that black tea, in the amounts studied, offered similar hydrating properties to water".

A 2014 study by Sophie Killer from Birmingham University and her colleagues in the UK measured not just urine volume, but also kidney function and the total amount of water in the body. The men in her study drank four cups of coffee per day. Even so, they showed no signs of dehydration.

What's the ultimate answer? Unfortunately, I'm left with the old adage "more research needs to be done". But at this stage, it seems as though your regular cuppa will neither dehydrate you, nor send you running to the bathroom like a tap.

1) Female 2) Getting older 3) Sun damage to skin 4) Certain medications → more ready bruising.

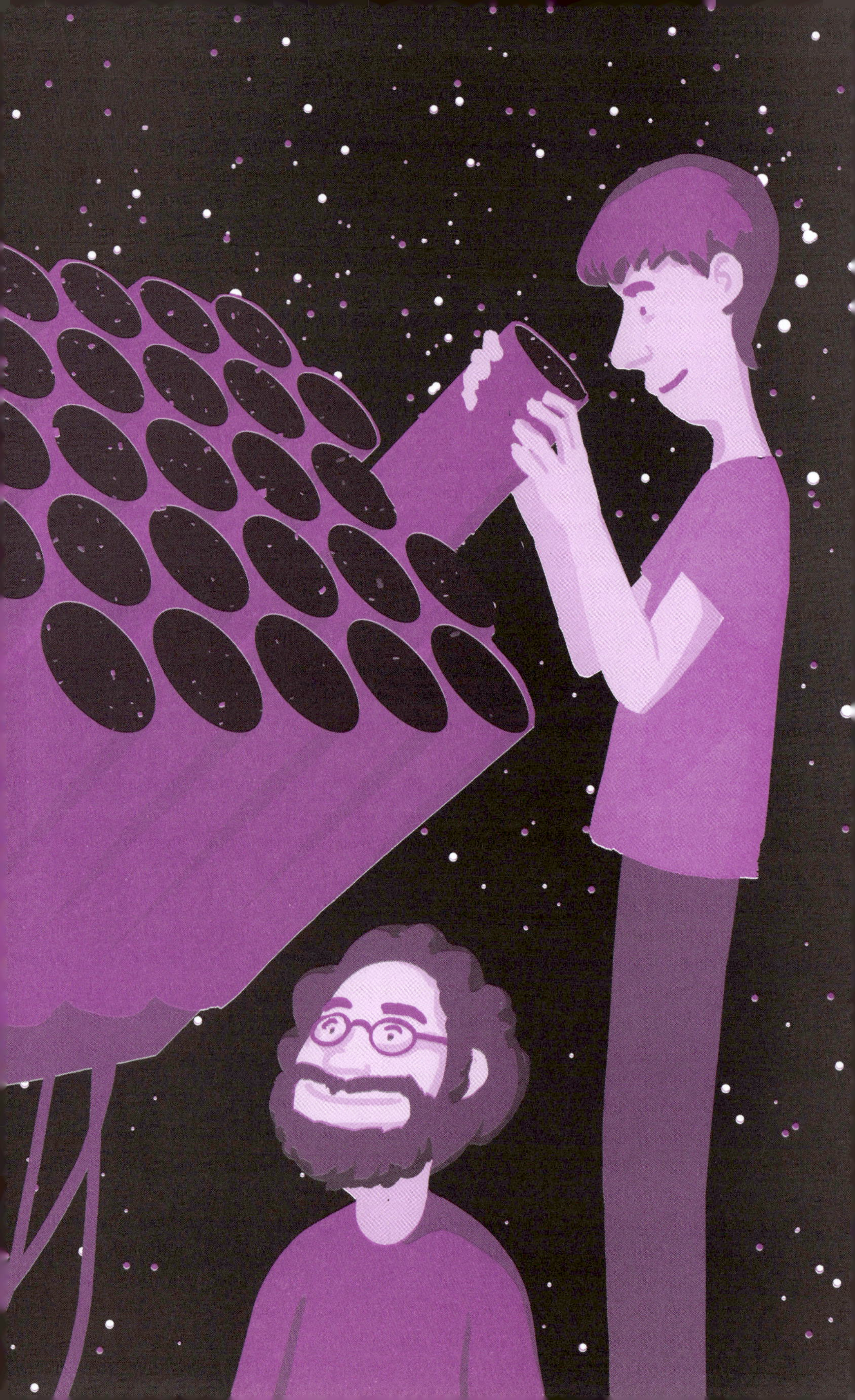

32

DRAGONFLY TELESCOPE

PROFESSIONAL ASTRONOMERS NEED NEW TELE-SCOPES LIKE FLOWERS NEED THE SUN. BUT TO DESIGN AND BUILD A NEW TELESCOPE, IT TAKES FORESTS OF PAPERWORK, BIG BUCKEROONIES – TENS OF MILLIONS OF DOLLARS MINIMUM – AND AT LEAST A DECADE. EXCEPT WHEN THE ASTRONOMERS COMBINE LATERAL THINKING, HOBBIES AND BEER.

Our Astronomer of Interest (aka Nature Photography Boy) was also lateral. His interest in cameras and lenses led to a cheap new top-notch telescope – and got us one step closer to solving the mystery of how galaxies spring into existence.

ANCIENT ASTRONOMY

Astronomy is one of the oldest sciences, going back to Ancient Egyptian times and further. Back then (like now), the professional Astronomers worked on the old motto, "I think, therefore I get paid."

They looked at the heavens and worked out a calendar. They could then tell the farmers when to plant their crops, so as to best benefit from the regular flooding of the Nile. In return, the Astronomers had regular, secure, paid government work.

HOW GALAXIES FORM

One major theory of how galaxies form involves Dark Matter. (I wrote about Dark Matter in my 34th book, *Game of Knowns*.)

This theory says that way, way back, over 13 billion years ago – in fact, just a few thousand years after the Big Bang – practically all the mass in the Universe was this mysterious Dark Matter. The Dark Matter began to clump together, thanks to Gravity, and began to shape itself into roughly spherical objects – which began to collapse inwards.

Various gases made of Regular Matter (such as hydrogen and helium) collected at the centres of these spheres, turning into the first stars – and the first galaxies. After a few billion years, the small galaxies merged with each other, eventually evolving into giant galaxies, like our own Milky Way.

But there's a catch with this Dark Matter Theory.

If this is how galaxies formed, there should be vast, messy debris fields floating around each big galaxy, left over from the creation process. We would expect to see random ejected stars, partially eaten halos of gas, bulges and streams of matter, and lots and lots of very faint dwarf galaxies.

But we couldn't find them. Had they been swept into a cosmic dust bin? Or is the theory wrong?

No. The problem was that the debris were very large and faint, and most telescopes don't work well on objects that are large and faint.

@DoctorKarl **How does tea and toast make you iron deficient?**

TELESCOPE: MIRROR VERSUS LENS

It turns out that most telescopes are inherently not very good at picking up the light from this leftover debris. It's soft and fuzzy and faint. And this is because the overwhelming majority of professional telescopes catch the incoming light with curved *mirrors*, not curved *lenses*. Curved mirrors are really good at stargazing and seeing small bright objects – which is the exact opposite of what we are looking for. If you think of a star as a blob of cream, and the leftover debris as dilute skim milk, you can understand why you'd need to use a different kind of telescope to see each of them.

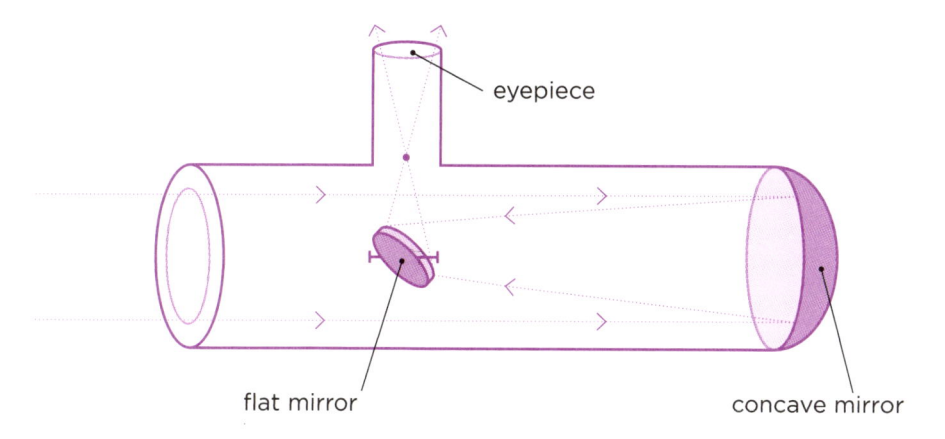

One good thing about curved mirrors is that you can make them really big (10 metres across). Over the last half-century, there have been tremendous advances and improvements in the performance of mirror-type telescopes (called reflective telescopes). In the specific field of gathering light from small bright objects, they're 100 times better than before. (I wrote about new telescopes in my 8th book, *Latest Great Moments in Science*.)

But when it comes to gathering faint light from large diffuse objects, mirror-type telescopes have not really improved over the last half-century. There are various technical reasons. First, the mirrors themselves cause scattering of the light, due to micro-roughness and

Tea has "tannin" → mop up iron. Bread has "phytates" → mop up iron. Can (e.g., widowed man who can't cook) → iron deficiency.

dust on the surface. This scattered light goes back into the beam – it both dims the view, and it also reduces contrast. Furthermore, the secondary mirror in mirror-telescopes creates a large obstruction in the light path. Finally, the supports that hold the secondary mirror cause diffraction and bending of the incoming light. (I wrote about this, and why you see stars as having points, in my 38th book, *Dr Karl's Short Back & Science*.)

So the ideal telescope for looking at large faint objects would have no mirrors and no obstructions to the incoming light. In other words, this Wonder Telescope would use lenses, not mirrors. With a lens telescope, any dirt or roughness on the glass surfaces would scatter light out of the beam. But apart from solar telescopes to look at the sun, professional Astronomers haven't used lens-type telescopes (refracting telescopes) for a century.

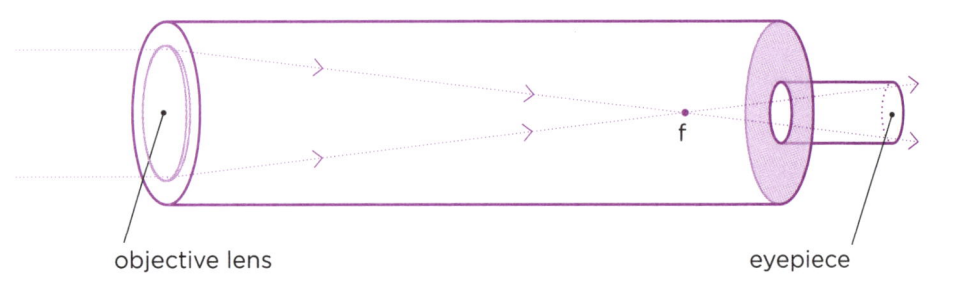

objective lens — eyepiece

THE WONDER LENS

The problem of how to find this leftover debris had long bothered two professional Astronomers, Roberto Abraham and Pieter van Dokkum.

It just so happened that back in 2011, they were shooting the breeze (as you do) about a quick and cheap way to find these large faint bits of debris.

Pieter van Dokkum was a keen nature photographer, and he suddenly realised that a recently released camera lens with a revolutionary new coating on the front might just be perfect for their needs. It was a Canon 400-millimetre $f/2.8$ SuperTelephoto lens, costing between

US$10,000 and $15,000. (I want one.) Canon engineers and scientists, at vast expense, had designed the coating with the specific purpose of reducing "flare", or scattering of light inside a lens.

This wondrous lens coating was called Subwavelength Structure Coating. It's actually countless tiny cones or pyramids (made from micro-crystalline alumina), on the front of the lens, all pointing outwards. These cones are microscopic – smaller than the wavelengths of visible light, hence the name Subwavelength. (See the story "Fly Eyes and Solar Panels" on page 83.) The Physics is complicated, but the end result is that less light is scattered inside the lens – so there's less of what the photographers call "ghosts" or "flares". It was first introduced in 2008 on a 24-millimetre $f/1.4$ wide-angle lens, and then on the 400-millimetre telephoto lens.

INSPIRATION?

In their paper, Abraham and van Dokkum write, "This project emerged as a result of a bet made by RGA and PGvD at the Mount Everest Nepalese Restaurant on Bloor Street in Toronto. We thank the United Breweries Group of Bangalore for providing us with inspiration on the night in question."

CHEAPER, BETTER, FASTER

And though it's not what this lens is designed for, it can pick up very faint objects.

In March 2012, Abraham and van Dokkum drove to the Mont-Mégantic Dark Sky Preserve in Quebec, Canada, with their Canon SuperTelephoto lens. Their off-the-shelf lens captured what other Astronomers had previously only got tantalising hints of. They saw a clear, but very faint, halo of diffuse matter surrounding the galaxy called M51.

Well, if one lens is good, surely three must better.

So they swiped the corporate credit card, got another two lenses, and built a special structure so that all three lenses were perfectly lined up.

1) If it's wet, keep it dry 2) If it's dry, keep it wet 3) If it's neither wet-nor-dry, or both wet-&-dry, use steroids.

ALMOST-PROFESSIONAL TELESCOPE

Even though it was designed as a camera lens, the optical performance of the Canon Super-Telephoto lens as a telescope is outstanding. Abraham and van Dokkum write that it is "capable of imaging extended structures to surface brightness levels . . . [that are] considerably deeper than the surface brightness levels of any existing wide-field telescope".

Very shortly afterwards, in September 2012, they got some results – yeah! By 2013, they were running eight lenses in parallel. In 2014, they published their findings that the galaxy called M101 (the Pinwheel Galaxy) had three previously undiscovered very faint dwarf galaxies orbiting around it.

DRAGONFLY ROCKS

The Astronomers, who admit they can't leave well enough alone, have since upgraded their telescope to 50 lenses. (The exact phrase in their paper is "One of the authors is congenitally unable to leave well enough alone, so . . .".) They can now get images in hours, not weeks. They didn't have to run through a few cubic metres of paperwork. Their corporate credit card bill has run up to about half a million dollars. But it's still a lot cheaper than spending tens of millions of dollars developing an entirely new telescope. It was also so much quicker, because the Canon scientists and engineers had done all the slow, expensive design work.

And why do they call their array of professional lenses the "Dragonfly Telephoto Array"? Well, with 50 commercial telephoto lenses, it looks like the eye of a dragonfly – not a mirror to the soul, but a lens to previously hidden galaxies. However, this is what Abraham and van Dokkum wrote in their 2014 paper about why they named their device the Dragonfly Telephoto Array:

Mind you, unlike professional telescopes, it "does not incorporate temperature compensation ... and focus changes become noticeable when the ambient temperature varies by as little as 1°C". But they worked around this.

They also found that the long body of the lens had a tendency to flex, so they added a three-point harness that grips the lens firmly near the front of the lens.

Even so, the camera lens (when set up in a multi-lens array) was a terrific bargain in cost, time and paperwork – while delivering roughly 10 times better performance than any other telescope.

1. We took inspiration from the compound eyes of insects when developing the central ideas behind the array;
2. The operational concepts that form the basis of the critical sub-wavelength nanostructure coatings were discovered by researchers investigating the unusually high transmittance of insect wings;
3. One of the authors really likes taking pictures of dragonflies.

› DATA › THEORY ›

The Dragonfly Telescope Array discovered 47 ultra-faint ghostly galaxies in the Coma Cluster (a crowded cluster of galaxies, about 300 million light years from Earth). They were as big in size as our Milky Way, but they were only a thousandth as bright.

So now we have this data (thank you, Dragonfly), it's time for coffee and theories.

One theory is that they are "failed giant galaxies". Early in their lives they would have fallen into a cluster of galaxies that cruelly stripped them of their "gas". No gas means no new stars being born.

A completely opposing theory is that they started as smaller galaxies that were spinning rapidly. The gas would have been thrown outwards, and any new stars were formed over truly enormous distances. So they now look faint and ghostly.

Which theory is correct? We need more data (and drumroll . . . more telescope observations).

No. Hair follicle → active (→ hair shaft thicker + more mm/day) → Reactive Oxidative Species → Dye injection unit dies → grey.

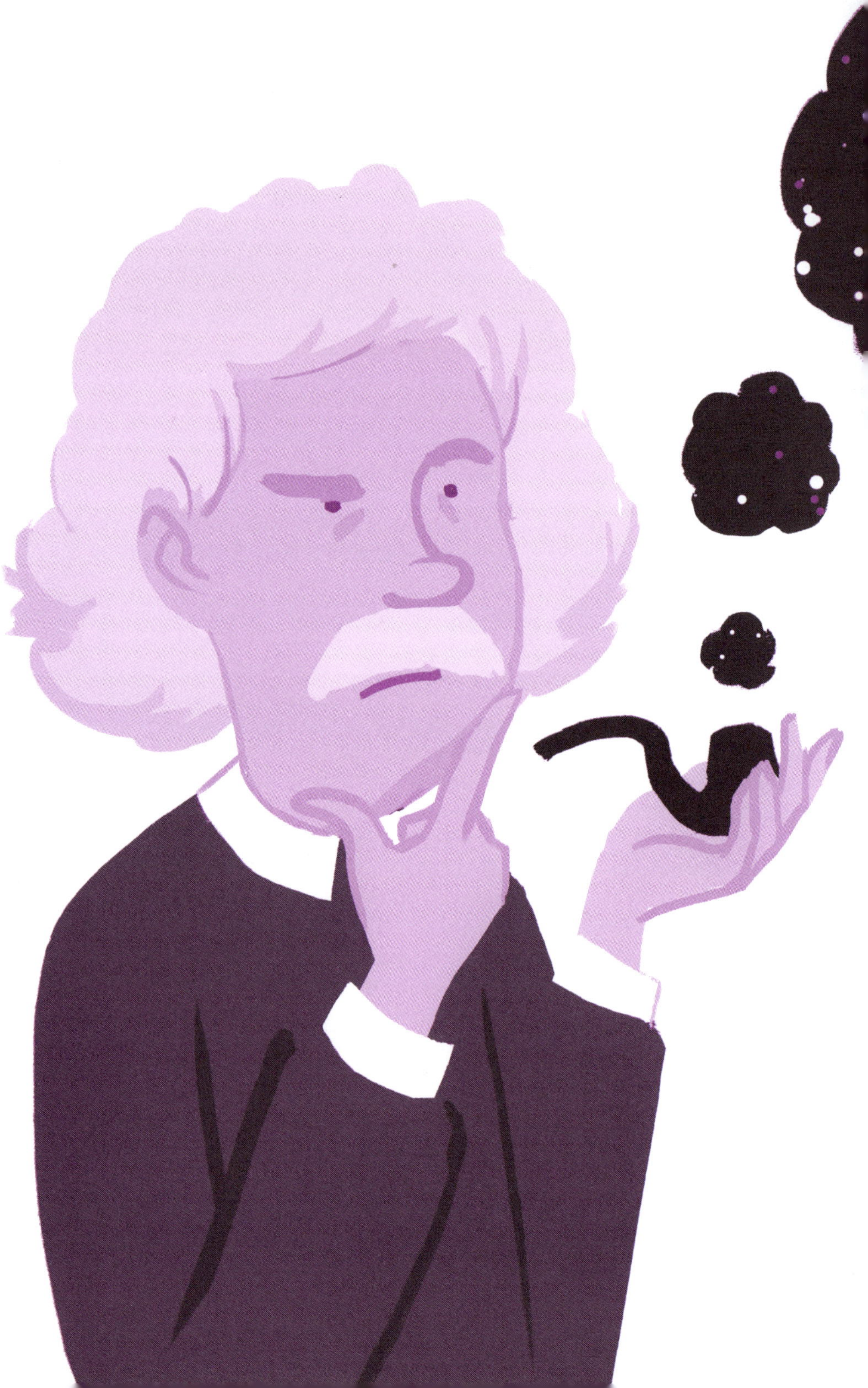

33 PHUNDA-MENTAL PHYSICS PROBLEMS

THE GREAT PHYSICIST ALBERT EINSTEIN SAID, "IF WE KNEW WHAT WE WERE DOING, IT WOULDN'T BE RESEARCH." FAIR ENOUGH. SCIENCE ADVANCES IN STEPS. FIRST, YOU START BY NOT KNOWING SOMETHING, THEN YOU THINK ABOUT IT, AND LATER, YOU UNDERSTAND IT.

Today, two of the biggest problems we have in Science relate to our understanding of Physics at a deep fundamental level. In each case, our measurements don't fit in with our theories – they simply don't make sense. Both the Higgs Field (Nobel Prize 2013) and Dark Energy (Nobel Prize 2011) are real. However, their measured values are much too small – astonishingly so.

This could turn out to be good in the long run. Almost certainly, once we solve the problem of why they don't match, a wondrous future will likely open for us.

CAN'T FOOL NATURE
Most scientists follow this aphorism: "I hold my theories on the tips of my fingers, so that the merest breath of fact will blow them away."

PROBLEM WITH THE SUN
This won't be the first (or the last) time that Science has to fine-tune its theories to fit the new evidence.

Around 1900, we didn't know how the Sun worked. When we solved that problem, we got something big in return.

Back then, the Astronomers had this Big Problem: "What is the power supply for the Sun?" They knew how much power the Sun emitted, and they knew the size of the Sun (it's enormous). Suppose the Sun was made entirely of coal, which was one of the most energy-rich fuels known back then. If the Sun were powered by coal, it would have burnt out after a few million years. And yet the Geologists were saying that the Earth was at least 10 or 20 million years old – and almost certainly, much older. It didn't make sense.

The Astronomers tried all kinds of workarounds. For example, they postulated that the effective mass of the Sun was much greater than we could see today, thanks to comets, asteroids and minor planets smashing into the Sun in the past. But even after adding in this extra mass, the numbers simply didn't match up.

The mysterious power-supply of the Sun turned out to be Radioactivity.

Very reasonably, you might say, "Well, this is nice theoretical knowledge, but what has it ever done for me?" Lots. On average, each Australian citizen can expect to have one nuclear medicine procedure (that uses a radioisotope for therapeutic or diagnostic reasons) at least once in their life. On top of that are all the X-rays you undergo. That nuclear medicine procedure, and X-rays, exist because we figured out that it was radioactivity that powered the Sun.

So let's whiz forward a century and look optimistically at two of the biggest Problems in Physics today.

HIGGS FIELD – 16 ZEROS WRONG

First, the Higgs Field. It creates the property we call "mass".

Your body is made of atoms. In turn, these atoms are made of particles called electrons and quarks. Here's the weird bit – inherently, these electrons and quarks have no mass. But when they interact with the Higgs Field, they get "mass" – and a Higgs boson is exchanged. A Nobel Prize was won for proving this.

The Higgs Field permeates all of space – and it's loaded with energy. Each cubic metre of the Higgs Field carries the amount of energy emitted by the Sun in 1000 years. That's a huge amount of energy.

But our theories tell us that the Higgs Field should be much stronger – about 10,000 trillion times stronger. We are wrong by 16 zeros. That's a lot of zeros.

Fidgeter? Dunno, but a truly dedicated fidgeter can burn up enough energy/day to run a marathon.

BEWARE OF DARK

Harry Cliff is a particle physicist at CERN (the European Organisation for Nuclear Research). He said during his December 2015 TED Talk, "Whenever you hear the word 'dark' in Physics, you should get very suspicious, because it probably means we don't know what we're talking about."

DARK ENERGY: 120 ZEROS WRONG

But if we have a Big Problem with the Higgs Field being too weak, we have a super-enormous Problem with Dark Energy being too weak.

What is Dark Energy? It was discovered in 1998, when astronomers found that the expansion of the Universe was actually speeding up. And what was forcing this speeding up of the expansion? Dark Energy. As with the Higgs Field, this discovery got a Nobel Prize for the Physicists.

But if you go into Quantum Field Theory and look at the Cosmological Constant, the Energy in the Vacuum and Dark Energy (and so on), you find that Dark Energy is too weak – by a factor of a trillion trillion trillion trillion trillion trillion trillion trillion trillion times. So our best estimate of Dark Energy is wrong by 120 zeros. That is much bigger than the biggest number in Astronomy.

It is also much, much bigger than the ratio in size between an atom and the entire known Universe – which is about 10 trillion trillion trillion times. (That's only 37 zeros.)

Here's another example of how wrong this estimate is. It's wrong by 1000 trillion trillion trillion (39 zeros), a factor of more than the number of atoms in the Universe.

Overall, you get the idea. The difference between what we measure, and what we predict, is so huge that a standard Physics textbook on

General Relativity properly called it "the worst theoretical prediction in the history of Physics".

THE WAY FORWARD

It could be that there's something very fundamental to both the Higgs Field and Dark Energy that we don't understand. Maybe, the answers lie in "supersymmetry" or "large extra dimensions". Or, perhaps we are simply putting together our pieces of knowledge and understanding the wrong way. And of course, there could be a dozen other explanations. In the early 21st Century, we simply don't know.

Now there's an old saying in Science, "It's not the Answer that gets you the Nobel Prize, it's the Question". Once you've got the Question, you just apply the necessary basics (such as brainpower, coffee, pizza, time, money, etc.) and do the Science for as long as it takes to get the answer.

We have two Big Problems and don't have the Answers – yet. But further down the line (today, tomorrow, or another century from now), we will. Just like the discovery of radioactivity in the Sun, light will be shed on these fundamental questions.

The Answers will not only generate a few Nobel Prizes, they'll give us essential gadgets and useful processes. And we currently can't even guess what these new gizmos will be.

Brain transplant = "you" have the personality/memories/emotions of the donor brain (if operation is possible).

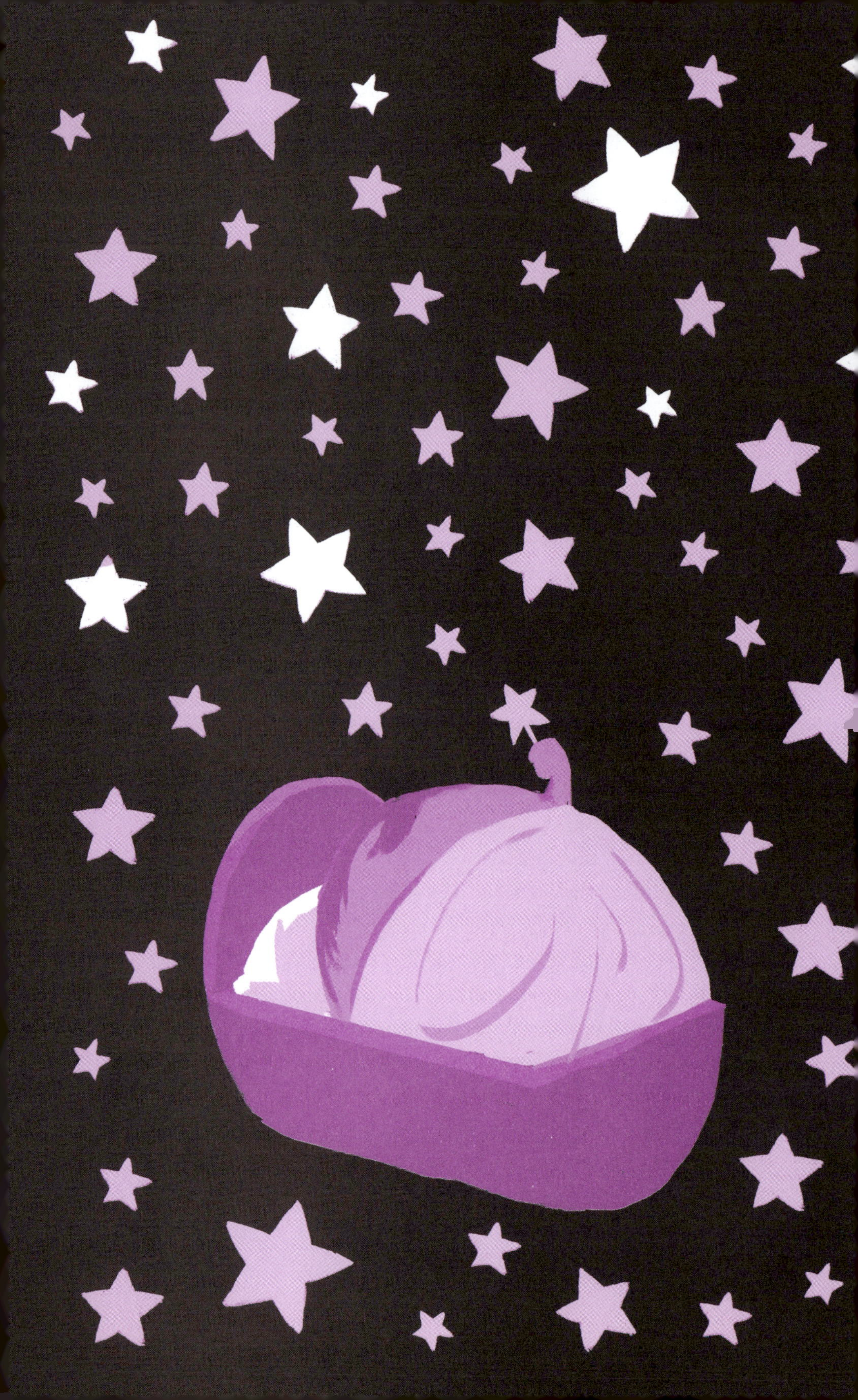

34

PLANET = A GIANT BABY

IT'S ALWAYS FUN WHEN A NEW BABY ARRIVES – TIME FOR FIZZY DRINKS AND CHOCOLATE!

But what if the baby is heavier than the planet Jupiter?

That's even better.

You see, the Astronomers have long worried about the mysterious process by which stars and planets are born from the apparent emptiness of space, inside a galaxy. They've had lots of theories, but not many facts. The arrival of this new giant baby planet gives astronomers a chance to test their hypotheses.

A STAR IS BORN

The earliest theories to explain the formation of solar systems were proposed by Emanuel Swedenborg in 1734, and further developed by the philosopher Immanuel Kant in 1755.

The basic hypothesis to explain where stars and planets come from is that atoms love each other very much, in a special way. This happens at special places in our galaxy.

Our home galaxy is the Milky Way. It has several hundred billion stars, and is about 100,000 light years across from one side to the other.

The emptiness of space between the stars is called the InterStellar Medium. It's very thin – as few as 100 particles in each cubic metre. The particles are about 90 per cent hydrogen atoms, with most of the rest being helium atoms. But in addition, there are tiny, tiny traces of heavier elements. These came from stars – both during their lives, and also at the ends of their lives, when they exploded and threw their own star stuff out into the InterStellar Medium.

But scattered about inside a galaxy are many Giant Molecular Clouds. There are about 6000 of them in our Milky Way. They can be a million times more dense than the InterStellar Medium – say, around 100 million particles per cubic metre. They can also be enormous – up to 100 light years across, and six million times the mass of our Sun. Between them, these Giant Molecular Clouds can carry half of the mass of the InterStellar Medium of a galaxy.

And inside these Giant Molecular Clouds are regions that are even more dense. They're called "Stellar Nurseries" or "Star-Forming Regions". As you can guess from the names, this is where stars are born.

@DoctorKarl Is Arsenic only poisonous because we've been accustomed to believe that?

WHERE STARS COME FROM

The first stage of the process deals with a star being born. It seems that it takes about a million years for a star to form.

In the Stellar Nursery, for whatever reason, there is one region that happens to be slightly denser than everything else around it. So it has more mass, and more gravity.

It collapses to form an object that is not quite a star – a so-called "protostar". After a few million years, it evolves further into a baby star, surrounded by a rotating disc of gas and dust that did not get sucked into the baby star. And when I say "baby", I mean "newborn", not "small". A baby star can range between a tenth of the mass of our Sun to 30 times more massive. (In general, in a solar system, the star will have well over 99 per cent of the total mass of that solar system.)

There are still some problems with our understanding of the process. Luckily, there are lots of stars out there. So we have seen different stages in the evolution from gas cloud to protostar to newborn star.

WHERE PLANETS COME FROM

The second stage is the formation of planets, orbiting around a star. It seems that it takes over 100 million years for planets to form.

This situation is more complex, and more difficult to understand. But the Astronomers have developed a rough theory – the Accretion Theory.

The theory starts with the gravity of a newborn star gathering gas and dust that didn't make it into the star. This matter gets formed into a dense, flattened disc that orbits in the plane of the equator of the star. (It's called a "Circumstellar Disc" or "Accretion Disc". It provides the material from which new planets are born.) It seems likely that within this disc, there are random locations that are more dense than other locations. These clumps of matter (planetismals) then sweep up more and more matter. Eventually, these planetismals turn into planets. (Almost certainly, magnetic fields are involved.) That is the rough theory. The trouble was that it had very few facts – astronomical observations – to back it up.

So this is the Big Question. What is the exact process as, over a few million years, the rotating disc of gas and dust evolves into a bunch of exoplanets (i.e. planets outside our solar system)? The problem is that there are so few exoplanets. By mid-2016, we had discovered only about 3500 exoplanets since the mid-1990s. All of them (except one) are fully formed. Quite simply, there was no solid data.

OUR FIRST BABY...

You guessed it. We have actually found an exoplanet, Lick-Calcium 15b, that is in the process of being born.

This planet is about 450 light years away. It's orbiting a newborn star called Lick-Calcium 15, which is only about two million years old. This baby planet is a few times heavier than Jupiter, and it's about three times further from its star than Jupiter is from our Sun (about 2.4 billion kilometres).

A team made up of astronomers from Stanford University, the University of Arizona and the University of Sydney made the discovery. One of the scientists, Professor Peter Tuthill from the University of Sydney, said, "we see the star surrounded by a disk of material, we see a gap in the disk where the material's missing, we see the planets that are in the gap, and we see material falling onto the planets." That's quite a jackpot to get from just one planet. Each of these individual events has been seen before – but this is the first time they have all been seen in a single planet orbiting a star. These are the birth pangs of a planet.

The newborn planet that is still growing, Lick-Calcium 15b, has a powerful magnetic field. This magnetic field actually accelerates the infalling material landing on the growing planet. As a result, this material is hot – very hot. It's around 10,000°C – roughly twice as hot as the surface of our Sun. But, being hot means that it emits lots of radiation, which our latest telescopes can detect.

With that kind of heat output, you would definitely want a baby shower...

@DoctorKarl What causes tingling and numbness in the legs?

FIRST ACTUAL IMAGE OF EXOPLANET

There are many ways to find a planet orbiting a distant star. (See "Other Solar Systems" in my 13th book, *Pigeon Poo, the Universe and Car Paint*.)

It was only on 13 November 2008 that researchers announced we had taken our first visible light direct snapshot of a planet orbiting another star.

The star is Fomalhaut, about 25 light years away. It's much hotter than our Sun, and about 16 times brighter. It's burning through its hydrogen at a furious rate, and will burn out in only one billion years (versus 10 billion for our Sun). It's quite young – only 200 million years old. Our Sun is about 4.7 billion years old.

Two separate Hubble Space Telescope photos in 2004 and 2006 showed an object that had moved in its orbit around Fomalhaut. This planet (with the unromantic name of "Fomalhaut b") takes about 872 years for a single orbit. This orbit lies about 10 times the distance of Saturn from our Sun.

Continual pressure on nerve for long time → numbness.

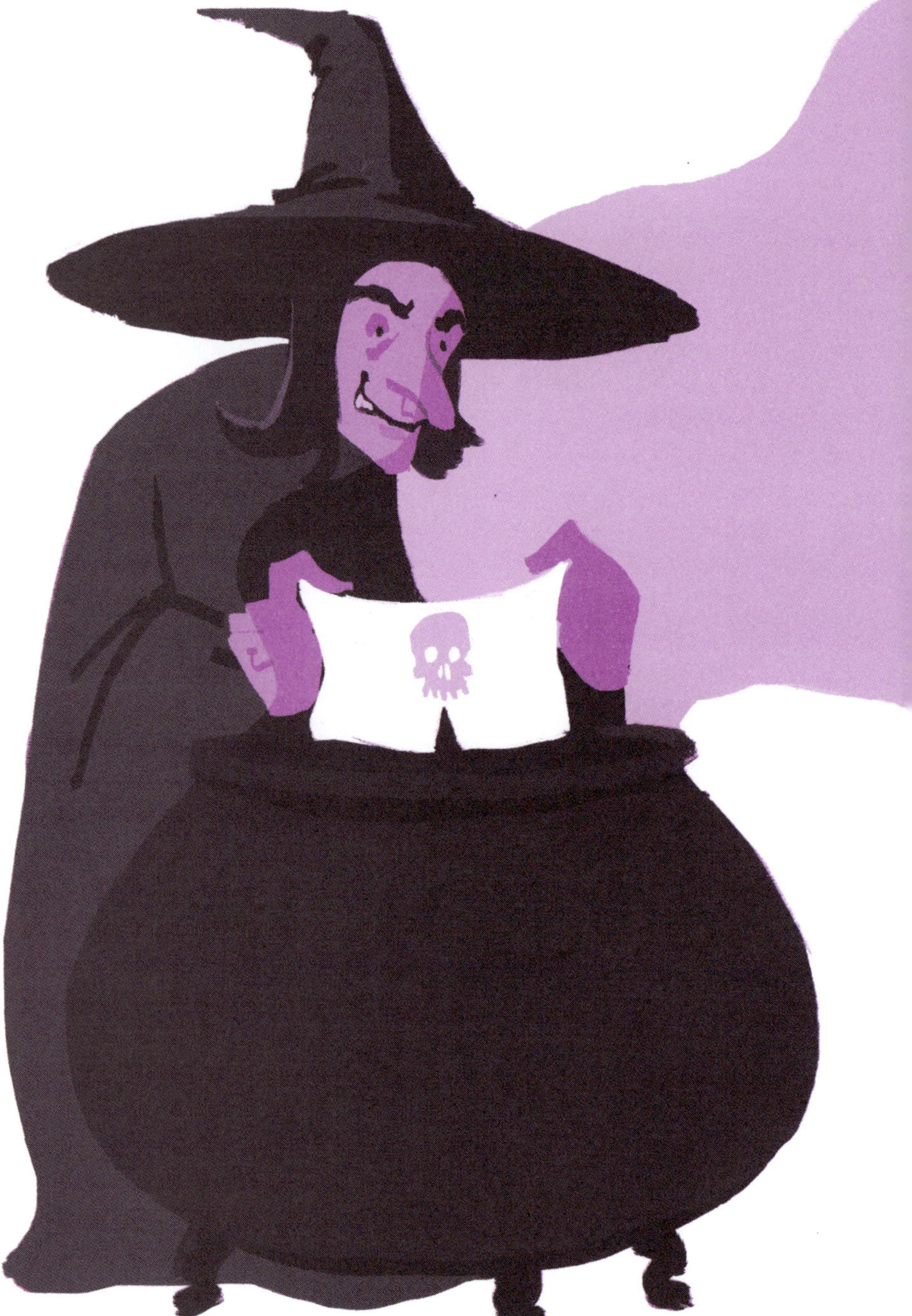

35

POISONED PANTS

ORAL POISONS HAVE LONG BEEN USED TO SEND ONE'S ENEMIES TO THE NETHERWORLD. CONTACT POISONS ARE DIFFERENT. THEY KILL NOT BY BEING SWALLOWED, BUT BY ENTERING THE BODY THROUGH THE SKIN.

They date back to Greek mythology. But they still are used today. Recently, paraquat was used in China, in the Case of the Poisoned Pants.

SHIRT OF NESSUS

The Greek hero Heracles died from a contact poison. In Roman mythology, and Disney movies, he's known as Hercules. (The precise historical details vary with the author, nationality and period of the myth, but here's a rough summary.)

In his Second Labour, Hercules, while still a demi-god, killed the dreaded Hydra. The Hydra was a multiheaded snake-like monster that lived in a lake. Both the breath and the blood of the Hydra were fatal upon contact, but Hercules was clever enough to cover his face with a mask. Never one to pass up a potential weapon, he dipped his arrows in the toxic blood of the Hydra.

Some time later, Hercules wanted to cross a river with his third wife, Deianira. Hercules said he would swim across. Nessus, a centaur and ferryman, offered to take Deianira across the river. After Hercules had swum across, Nessus tried to rape Deianira. Hercules saw this, and shot Nessus with the poisoned arrow from the opposite shore. As he lay dying, Nessus plotted revenge. Pretending to make up for his terrible actions, he told Deianira that if she soaked Hercules' clothes in his centaur blood, its special properties would make Hercules faithful to her. So Deianira naïvely collected the now-contaminated blood of Nessus. (The poison of the Hydra was still active in the centaur's blood.)

On a later occasion, Deianira was – with good reason – apprehensive of Hercules' faithfulness. She soaked one of his shirts in the blood of Nessus, and instructed one of the servants of Hercules to bring it to him. He put the shirt on, and immediately began to suffer in agony. He tried to tear the shirt off, but it stuck to his skin and ripped his flesh off his bones. Realising he was doomed, Hercules started a funeral pyre, and voluntarily threw himself upon it. He died, became a full god – and married his fourth wife.

So as a literary metaphor, the Shirt (or Tunic) of Nessus is a fatal present, or a source of misfortune.

@DoctorKarl Could you have ocean waves without a moon?

PARAQUAT TOXICITY

Paraquat is a highly toxic herbicide, and there is no known antidote. There have been 18 deaths from it in Australia since 2000.

Because the alveolar epithelial cells in the lungs selectively accumulate paraquat, its toxic effects first show in the lung tissue. It quickly develops into pulmonary oedema, followed by fibrosis. If you accidentally swig some paraquat, and then immediately spit it out, you will still get lung fibrosis. In the UK in 1992, an agricultural worker "died after being splashed in the face with paraquat after he dropped an open container".

Paraquat causes severe problems – both short-term and long-term. These include respiratory failure, severe dermatitis, kidney failure, second-degree burns, skin cancer and even Parkinson's disease.

It is dangerous for four reasons – it's very toxic, there is no antidote, it's readily available and is quite cheap. This leads to it being widely used as a suicide agent in poor countries. Death follows slowly and miserably, taking from several days to weeks after ingestion.

In poor countries, once the sales were restricted or controlled, death rates from paraquat poisoning dropped to one third their previous levels within three years.

In some parts of the world, three so-called "alerting agents" have been added to the paraquat herbicide. They are a stenching agent (to smell bad), an emetic (to induce vomiting) and a blue dye (to avoid confusion with dark beverages such as cola or coffee).

PARAQUAT 101

However, excluding the toxic blood of the mythical Hydra, human skin is quite effective at blocking chemicals from passing through it. Skin has a water-based layer that stops fats, and a fat-based layer that stops water-soluble chemicals. But if something dissolves in both oil and water, then it can slip through the skin – and paraquat can do this.

Paraquat does not pass immediately through unbroken skin. But it's an irritant and damages the skin – and then paraquat can enter through the damaged flesh.

Yes. WAVES are caused by the "fetch" – the wind blowing over long distance of water. Moon ($2/3$) and Sun ($1/3$) make TIDES.

PARAQUAT HISTORY

Paraquat was first made in 1882. However it took until 1955 before its herbicidal properties were discovered. It was first produced commercially in the early 1960s.

Worldwide, it was the most widely used herbicide until the introduction of glyphosate (Roundup).

Paraquat is a non-selective weed killer – "non-selective" meaning that it attacks both plants and animals.

It interferes with the transport of electrons – a process essential to life. It also creates the so-called "Reactive Oxidative Species", which chemically damage tissue they come in contact with. It kills plants by interfering with photosynthesis.

Paraquat as a herbicide is often used on broad-leaved plants, but is less effective on plants with deep roots and does not harm mature bark. So it's used to control weeds in fruit orchards and plantations of crops such as bananas, coconuts, coffee, rubber, etc.

It binds to non-sandy soil and tends to stay in the soil. Unfortunately, it degrades quite slowly at about 5 to 10 per cent per year.

PARAQUAT POISONING

Paraquat needs utmost care in handling and using it. Most wealthy countries have banned it, or allow its use only under stringent conditions. However, in some poor countries in which it is used, protective gloves cost a day's wages, and so paraquat is used without the proper safety procedures. Furthermore, the workers who spray it are usually not trained in how to use it, and medical facilities are few and far between. And so in many poor countries, paraquat is the most common poisoning agent, with a death rate around 60 to 70 per cent.

@DoctorKarl **Is it possible for the Hadron Collider to blast out another universe?**

For humans, the LD_{50} (the Lethal Dose that will kill 50 per cent of recipients) is 35 milligrams per kilogram of the body mass of the recipient – a few grams, or less than a teaspoon of paraquat weedkiller. According to a World Health Organization report in 2000, worldwide there were some 2 million cases of paraquat poisoning each year – with a 10 per cent death rate. Of those who died, about 90 per cent were deliberate poisonings (mostly suicide).

GETTING BACK TO THE POISONED PANTS

In May 2016, Mrs Zhang from Hangzhou in Zhejiang province, China was still angry with her husband after a row. She soaked his underpants in paraquat, let them dry, and gave them to him so he could wear them for their daughter's wedding.

Mr Zhang was taken to hospital when his genitals began rotting, and he had difficulty in breathing. Luckily he recovered after three weeks in hospital. So he survived, but his genitals are another story.

Doesn't have enough energy. There's more energy in incoming cosmic rays.

216 > The Doctor

@DoctorKarl How can a fly (top speed ~10 kph) keep up, when flying inside a speeding car at 100 kph?

HETEROPATERNAL SUPERFECUNDATION > MILKY WAY

Hercules (known by the Greeks as Heracles) had a complicated love-hate relationship with his stepmother, Hera. Hera was the wife of Zeus, the King of the Gods. Zeus, unfortunately, had many affairs with other women, and many "illegitimate" children.

In the case of Hercules, Zeus made love to the mortal woman Alcmene by misusing his godly powers to disguise himself as her husband, Amphitryon, who was away at war. By an amazing coincidence, Amphitryon returned later the same night – and also made love to her. Alcmene became pregnant to both Zeus and her husband, giving us the rarely used phrase "heteropaternal superfecundation" – where a woman delivers twins who have different fathers.

As you might imagine, Hera did not like this general state of affairs. More specifically, she deeply disliked Hercules, and continually tried to make his life miserable.

However, in a twist worthy of a telenovela, Athena (the protectress of heroes) brought Hercules to Hera. Hera did not recognise Hercules, and to keep him alive, wet-nursed him. But even as a baby, Hercules was very strong. He suckled Hera so powerfully that she cried out in pain and pushed him away. Her milk sprayed across the heavens – and coagulated into the Milky Way. But Hercules had already drunk enough of her godly milk to become a demi-god.

And that's how heteropaternal superfecundation and Hercules gave us the Milky Way.

The fly's "inertial frame of reference" is the car, not the road.

36 I, VOMIT

Even without studying the language, there is probably one Latin word you know – "vomitorium". And "everyone knows" the vomitorium was where, back in debauched Roman times, gluttonous eaters would go to vomit.

Sorry, but vomiting is not what you're supposed to do in a vomitorium.

THE MYTH

So why do so many of us think that's true? Probably we picked up incorrect knowledge about the vomitorium at school from teachers or friends, or afterwards from one's friends. (For me, it was both sources.)

Various reasons were given for visiting the vomitorium.

Even though there's nothing written, the myth suggests that the Romans visited the vomitorium between courses at the banquet to make more room in their tummy. Or, they vomited as a sign to others of their fabulous wealth. They didn't eat for nutrition, but as a sign of conspicuous consumption. Or, they vomited to keep their weight down.

Regardless of why, in popular parlance, they went to the vomitorium to vomit.

THE TRUTH

The word "vomitorium" does indeed come from a Latin root meaning "to spew forth". But it doesn't refer to the contents of one's stomach. A vomitorium is a passage or opening in a theatre (or amphitheatre), leading to or from the seats, through which the audience would pass. These passages would be big enough to rapidly disgorge the audience – in the case of a fire, or even just another show immediately afterwards.

@DoctorKarl Do bodies decompose in Space?

The Coliseum in Rome had 76 vomitoriums (technically "vomitoria") for the common spectators to enter or leave. There were another four vomitoriums dedicated for the exclusive use of the Imperial family. As a result of having these 80 spacious entrances/exits, 50,000 people could enter or leave the Coliseum within 15 minutes.

VOMITORIUM IN LITERATURE

The word "vomitorium" was first used in the fourth century AD by the Roman writer Macrobius. He wrote about the amphitheatre passageways that could "disgorge" the audience to and from their seats. It has long been used by architects, builders and theatre folk in its correct context.

In the theatre, the word is abbreviated to "vom". So a director will coach an actor and tell them that they should leave via the "stage right vom".

The first incorrect use of vomitorium seems to have been by Aldous Huxley in 1923, in his comic novel *Antic Hay*. He wrote, "There strode in, like a Goth into the elegant marble vomitorium of Petronius Arbiter, a haggard and dishevelled person". The urban historian Lewis Mumford made a similar mistake in his 1961 work *The City in History*. He wrote that greedy Roman eaters had "thrown up the contents of their stomach in order to return to their couches". However, he also wrote that only later did the word get linked to entrances/exits at the stadium.

Some modern theatres incorporate "voms". In addition to allowing the movement of the audience, they are often used by the actors to enter and exit. The Cockpit Theatre in London was built in the 1960s as a theatre-in-the-round (actually a square). It has four voms, one between each bank of raised seating. In New York, the Circle in the Square Theatre on Broadway also has several voms.

So if you do vomit in the vomitorium, don't expect anyone to applaud.

No. Bacteria in gut would still exist for short time. Their activity would cease once temp dropped enough.

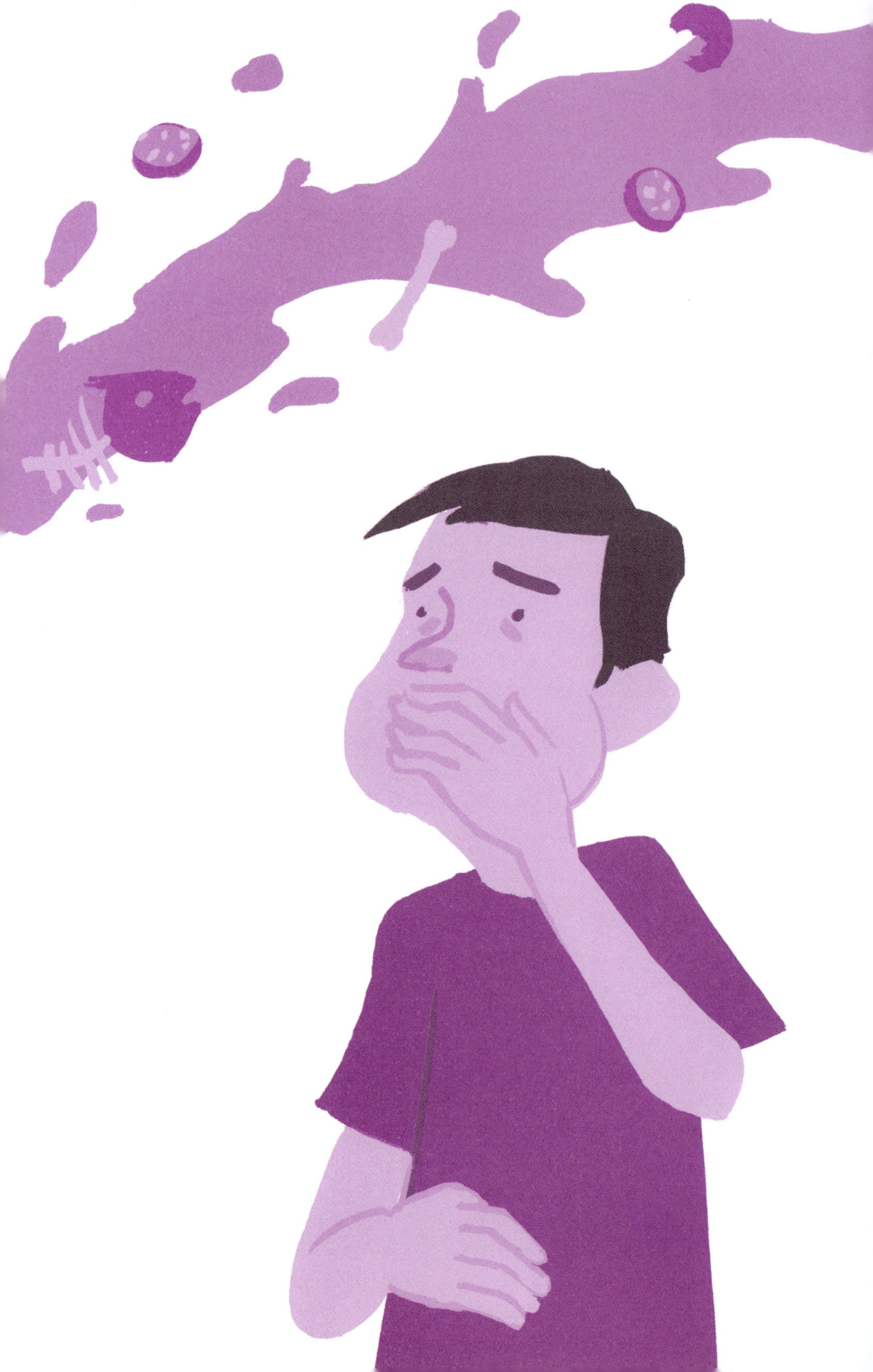

37

VOMITING MACHINE

DISGUST IS A FEELING OF STRONG REVULSION OR DISAPPROVAL – AND VOMITING IS ONE OF THE MOST DISGUSTING EXPERIENCES.

In fact, even seeing another person vomit can make an otherwise well bystander start vomiting.

So why would a medical doctor work with biological scientists and engineers in North Carolina to build a vomiting machine – in fact, a projectile vomiting machine? No, not because they were "sick", or wanted to make other people vomit. They wanted to understand how Noroviruses from infected humans spread through the air.

VOMITING 101

Vomiting is an involuntary and forceful expulsion of whatever happens to be in your stomach out through your mouth – and sometimes your nose.

There seems to be a Vomiting Centre in your brain (the Area Postrema). The Area Postrema lies outside the Blood–Brain Barrier, so chemicals in the general bloodstream can affect it. (I've written a story on the Blood–Brain Barrier in my 34th book, *Game of Knowns*.) Various nerves connect to it, such as Cranial Nerve VIII (which deals with balance) and Cranial Nerve X (which among other things, deals with your gut). The Vomiting Centre also has receptors for chemicals such as dopamine, serotonin, opiates and so on.

Believe it or not, vomiting is a beautifully choreographed act. (But that doesn't necessarily make it beautiful.)

In the lead-up to the actual act of vomiting, your lungs will take a deep breath – this minimises the chance of vomit getting into your lungs. You'll also start salivating – this helps protect your tooth enamel from the stomach acids that can cause dental erosion.

You then both generate higher pressure in the abdomen (as your tummy muscles contract) and lower pressure in the chest (as you try to breathe in against a closed larynx). These two effects combine to propel the contents of the stomach into your oesophagus – and through your mouth, and out.

Sometimes, people can vomit so forcefully and for so long that their tummy muscles get sore.

CAUSES OF VOMITING

There are many causes of vomiting.

With regard to the gut, vomiting can be set off by gastritis, food poisoning, overeating, food allergies, lactose intolerance in children, and many other conditions.

Considering the sensory system and the brain, vomiting can be caused by motion sickness, concussion, migraine, brain cancers and more.

Other causes include metabolic disturbances (altered levels of calcium, urea, glucose, etc.), pregnancy, drugs (such as opiates, alcohol, etc.) and yes, gastroenteritis from infectious agents such as Norovirus.

NOROVIRUS = FOOD POISONING

There are several different viruses in the Norovirus family, ranging from 23 to 40 nanometres in diameter. They are tough, and can survive temperatures from freezing to 60°C. Because they are a single-stranded RNA virus, once inside a cell they will quickly set up "replication complexes", which will each turn out hundreds more copies of the infecting Norovirus.

In the UK, it's sometimes called the Winter Vomiting Bug. It's the most common cause of viral gastroenteritis in humans. The gastroenteritis usually develops 12–48 hours after the Novovirus enters the body.

According to the Centers for Disease Control in the USA, Norovirus is responsible for about 70 per cent of reported US infectious gut disease outbreaks – and 46 per cent of hospitalisations, and 86 per cent of deaths associated with those gut disease outbreaks. These outbreaks happened mostly in healthcare facilities (64 per cent), but also in food service establishments (e.g. restaurants and banquets, 15 per cent). But of course, the location that gets the big publicity in the media is the cruise ship, carrying a few thousand passengers.

The effects of Norovirus (which was previously called Norwalk Virus and Norwalk Agent) are quite nasty.

Different. CO binds ~500 x better than O_2 to haemoglobin (and irreversibly in short term).

Each year worldwide, it infects about a quarter of a billion people, and kills 200,000 of them.

But in most cases, people recover fully within a few days – even though the symptoms are unpleasant. These symptoms include nausea, projectile vomiting, tummy pain, watery diarrhoea, headache and a low-grade fever. Symptoms generally last for 24–72 hours.

Norovirus is classically associated with gastroenteritis outbreaks on cruise ships, aircraft, hotels, schools and public gatherings.

It is spread by contaminated food or water, direct person-to-person contact, and yes, through the air – which is where vomiting comes in.

NOROVIRUS AND ALCOHOL SWABS

Norovirus can survive for up to 12 days on contaminated clothing, weeks on hard surfaces, and months or perhaps years in contaminated still water.

So it can be a problem where large numbers of people congregate, such as big corporations, cruise ships, and so on.

Often, antiseptic alcohol swabs or pump packs are made available to clean hands in high-traffic areas, such as elevators and entrances to dining rooms.

Unfortunately, alcohol is not particularly effective against Norovirus. Noroviruses do not have a surrounding envelope of fat, so they are relatively resistant to alcohol. The best technique to avoid hand-to-mouth transmission is to wash your hands thoroughly with soap for about two minutes. The surfactants (detergents) in the soap lift up the dirt on your hands and get rid of the virus – but you do need lots of friction (hence the two minutes of washing).

NOROVIRUS FLOATS IN AIR

Here's a typical example of how Norovirus can travel so very easily through the air.

In December 1998, 126 people were at a function eating in a restaurant, which had six tables. One person had already been infected with Norovirus. During the meal, the infected person quietly vomited onto the floor. (Yes, disgusting.) The staff quickly and efficiently cleaned it up. People very close politely pretended that nothing untoward had happened – and people further away didn't even know. Everybody continued eating.

But from that one vomit, Norovirus particles had been unleashed, and travelled through the air to other people in the restaurant. Over the next few days, 52 of the 126 partygoers got sick.

The infection rate was related to the distance from the Vomit Epicentre. At the same table as the unfortunate vomiter, over 90 per cent got infected. At nearby tables, the infection rate was 50–70 per cent. And way on the other side of the restaurant, 25 per cent got Norovirus.

VOMIT MACHINE

So from anecdotes like this, it was pretty obvious that Norovirus particles could infect people, through the air, from a considerable distance.

But until the North Carolina team got together to build a projectile vomiting machine, nobody had actually measured how much virus was released during a typical vomiting episode. This was why they built a one-quarter scale Simulated Vomiting Device. (For the follow-up studies, they should try a full-scale Vomit Machine, adding "solids" to the simulated vomit, etc.)

Norovirus is pretty nasty, so the team loaded up their Simulated Vomiting Device with a well-studied and relatively harmless virus known as Bacteriophage MS2. (I wrote about bacteriophages, and how

Nature is bloody in tooth and claw. They get rapidly eaten by hungry living creatures.

they kill half the bacteria on Earth every two days, in my 38th book, *Dr Karl's Short Back & Science*.) MS2 is similar to Norovirus in shape and size. The researchers' device had a simple hand pump to pressurise it, various valves to let the pressure build up and be suddenly released, and a clay face mask wrapped around the output pipe giving a suitable expression of misery.

(OK, besides adding an artistic flair, the face mask acted as a weight. Its other purpose was to impart the same curve to the plastic pipe that the human throat has while vomiting. The advantage of having your throat curved towards the ground during vomiting is that this reduces the chances of inhaling vomit into your lungs.)

Now, vomit varies in consistency. So the experimenters tried to be very thorough with their experimenting. They made vomits of different consistencies – ranging from watery artificial saliva, to thick gluggy vanilla instant pudding.

In a full-scale human vomit, the maximum volume is around 800 millilitres, including some 50–200 millilitres of air. This air helps turn some of the wet vomit into tiny aerosols floating in the air.

The team found that the number of virus particles released from the mouth of their Simulated Vomiting Device ranged from 36 right up to 13,000. And how many Norovirus particles do you need to get infected? Just 20.

That's a (w)retchedly small amount to trigger such dreadful vomiting...

@DoctorKarl **Why can you **see** a light in the distance, even when it's not illuminating your body?**

Vomiting Machine < **227**

**In optimal conditions, your retinas can respond to just a few photons →
signal to brain. Same few photons over body → can't be seen.**

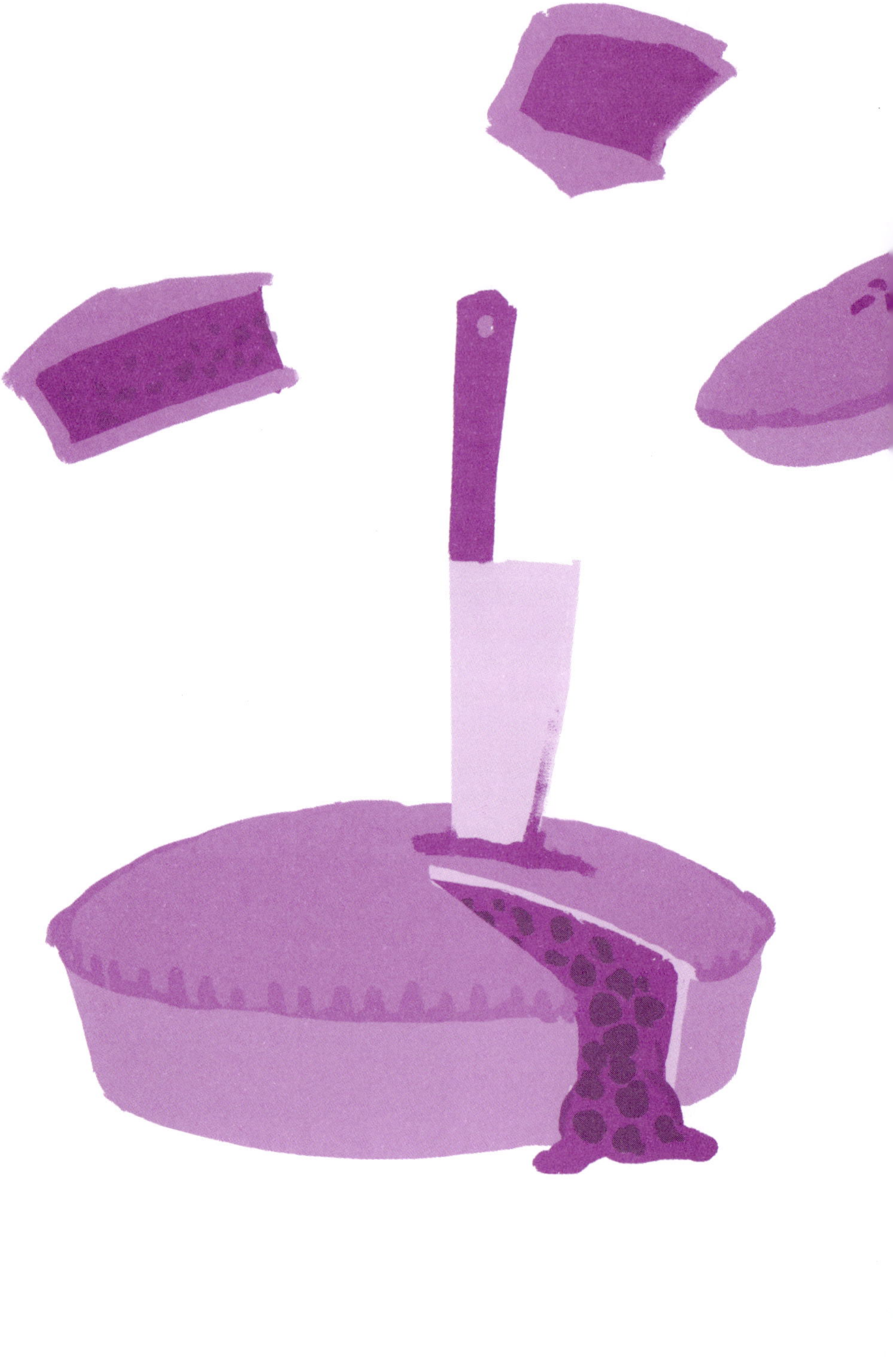

38

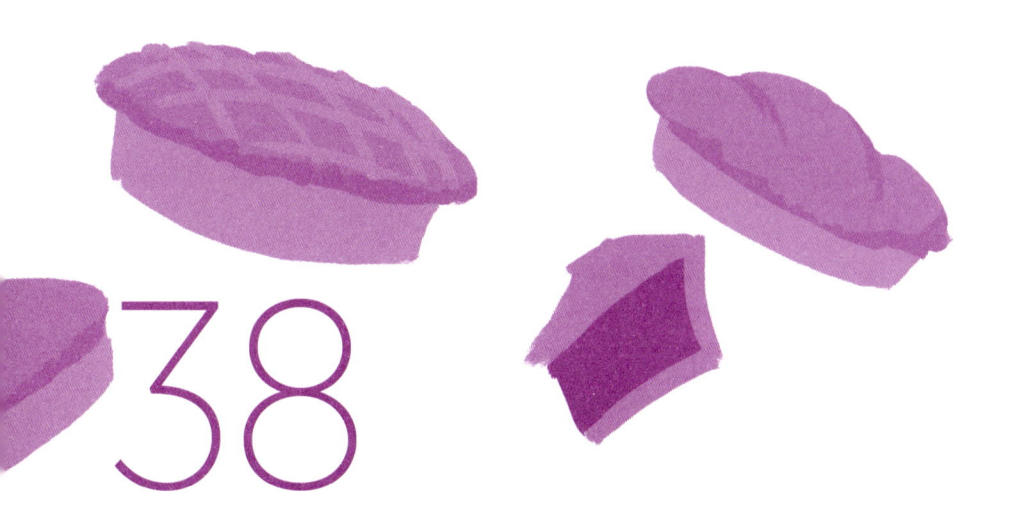

PLACES OF PI

IN THE LAND OF FOOD, A PIE IS SOMETHING DELICIOUS FROM AN OVEN. BUT IN THE LAND OF MATHEMATICS, PI IS A VERY SPECIAL NUMBER – AND IN ITS OWN WAY, QUITE TASTY.

Some special people have memorised Pi (π) to tens of thousands of decimal places. But we actually need fewer than a handful of digits for everyday use. And did you know that you can work out Pi by firing a shotgun at a target, or throwing sticks onto a tiled floor?

PI 101

Pi underlies much of what we know about the Universe. This is because many objects, and their movements, are pretty close to spherical or circular. The shapes of stars, the orbits of planets, the movement of radiation and energy – they are all roughly circular or spherical.

Pi appears in the spiral of our DNA, the rainbow, the Bell Curve that describes things such as distribution of IQs and death rates, and colours and music.

Pi appears in both General Relativity, and Quantum Physics.

Pi is the Magical Number for a circle. It's the ratio of the circle's circumference to its diameter. For most uses, you can get away with Pi as being equal to 3.1416 – that's four decimal places. So in the "real world", one roll of a wheel will cover just over three times its diameter.

ORIGIN OF PI

Where did the name "Pi" come from? Well, in the Greek language, "peri" means "around", while "meter" means "measure". So "peri-metros" referred to the "circumference" of a circle. The first letter in that Greek word is the letter "π", or "Pi".

The idea that there was a fixed ratio between a circle's diameter and its circumference was known way back in ancient Egypt and Babylon. By around 1800 BC, these cultures each had values for π that were accurate to within 1 per cent. The Babylonians treated π as though it were the fraction $22/8$, which equals 3.125 (just under 3.1416), while the Egyptians preferred the approximation $(16/9)^2$. This roughly equals 3.1605 (just over 3.1416).

The English mathematician William Oughtred was probably the first to use the symbol π to represent the circumference–diameter ratio, back in the year 1647. But Leonard Euler, an awesomely brilliant Swiss mathematician, really popularised π after he used it in one of his books about a century later.

WORKING OUT PI WITH GEOMETRY

This brings me to a major fact about π. This lovely number is not equal to any simple fraction of one number divided by another number.

You're fairly close with $22/7$. You get even closer with $355/113$. But π is not exactly equal to any fraction.

EASY SIX DECIMALS OF PI

Here's a trick to remember $335/113$.

Start with the sequence 113355 – fairly easy to remember. Split it in two ($113/355$) and turn it upside down ($335/113$).

Next you can work out the fraction. The answer is the same as π for the first 6 decimals – 3.141 592 – and then it diverges.

So, how do you work out π?

Sure, you can actually measure (with a tape or piece of string) the circumference and diameter of a circle. The problem is that you can never measure anything exactly. There are always errors in measurement..

You can get a lot closer with Geometry. Draw two polygons – one just inside a circle, the other just outside. The circle is sandwiched between the two polygons.

Archimedes did this around 250 BC. A polygon with 6 sides is a bit bumpy. But one with 12 sides is closer to the smooth shape of a circle. He then did the maths for polygons with 24, 48 and finally 96 sides.

(In the drawing on the next page, in the left hand column the purple polygons are just inside the circle. In the right column, they are just outside the circle.)

For each polygon with 96 sides, he worked out its ratio of the circumference to the diameter. He came up with π for a circle as being somewhere between $223/71$ (for the outside polygon) and $22/7$ (for the

---

Fireplace burning 1 day emits more particle pollution than car running 15,000 km. That's why a single fireplace can fill valley with smoke.

PI = LUDOLPHIAN NUMBER

In 1596, a German-Dutch mathematician, Ludolph van Ceulen, worked out π to 35 decimal places (that's one part in one hundred million billion billion billion).

He was so pleased with himself that he made sure that all those digits were engraved on his tombstone, at the Ladies' Church in Leiden, in the Netherlands. His remains and his beloved tombstone were later

inside polygon). The Ancient Greeks hadn't invented decimals, so he had to do all his working out with fractions – which made his job harder. In modern terms, Archimedes had showed that π was somewhere between 3.1408 and 3.1429 – a pretty good result.

Of course, if you went for more than 96 sides, you could get better estimates. By the late 1500s, other mathematicians had achieved accuracies of 15 and 20 decimal places in the value of π. The best accuracy for π they ever reached by using polygons (with huge numbers of sides) was 39 digits. But this method took a massive amount of work.

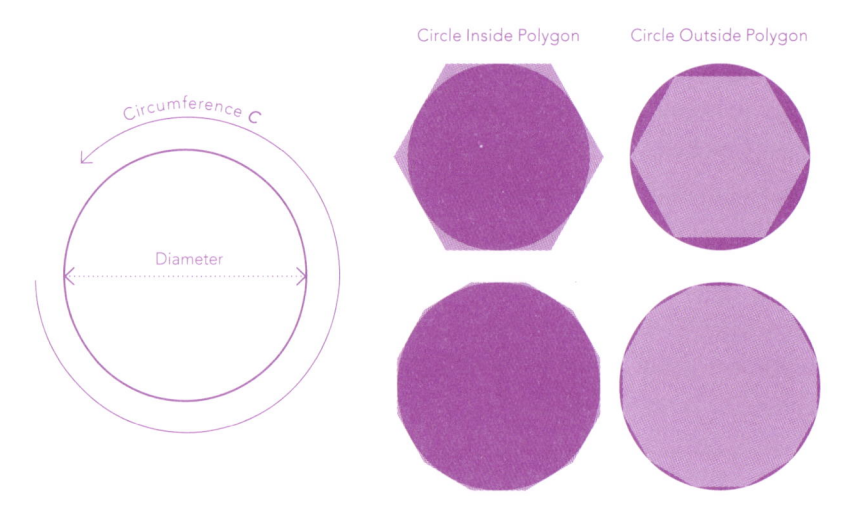

shifted to Peter's Church in Leiden, to a special graveyard for professors. Unfortunately, the tombstone then vanished for several centuries, before being found in 2000.

In his honour, Germans called π the "Ludolphian Number" or the "Ludolphine Number" until early in the 20th century.

WORKING OUT PI WITH INFINITE SERIES

No, the best way to get lots and lots of digits of π lay in the development of Infinite Series in the 1500s and 1600s.

Now Infinite Series can be really weird.

You would normally think that if you added up an infinite number of terms, the result would be infinite. Sure, the Infinite Series

$$1 + 2 + 3 + 4 \ldots$$

will eventually add up to infinity.

But if you choose your terms, the sum of an infinite number of terms is *not* infinite – it can be finite. (This is definitely weird. You do an infinite number of additions and you get a finite result?! Don't worry, the Universe is full of weird stuff. Maybe it's just that our brains aren't weird enough to understand...)

Consider the series

$$1 - 1/3 + 1/5 - 1/7 + 1/9 - 1/11 + 1/13 \ldots$$

German polymath Gottfried Wilhelm Leibniz came up with this series in 1674, while thinking about circles. (It has been called one of the most

---

Most probably not. 93 per cent of heat from Global Warming (as trapped by CO_2, etc.) has gone → warming oceans.

beautiful mathematical discoveries of the 17th century. However, it had been independently discovered around 1500 by the Indian mathematician Nilakantha Somayaji.) This Infinite Series adds up to $\pi/4$ – which is definitely a finite number. Way back in 1699, the English mathematician Abraham Sharp used this particular infinite series to push π up to 71 digits. But that calculation took him an enormous amount of time.

Since then, the mathematicians have devised better Infinite Series – better in the sense that they get closer to π in a smaller number of steps. Furthermore, our scientists and engineers have given us increasingly fast computers, which make adding up numbers much quicker.

So we worked out π to 10,000 digits in 1958, a million digits in 1973, and now we're up in the trillions.

And certain humans have been able to memorise up to about 70,000 places of π.

HOW MUCH PI DO YOU NEED?

So how many places of π do we need in our modern society? (I guess some of you might be quietly wondering if we need any at all.)

Well, if we assume that π has a value 3.142, that gives us accuracy better than 1 part per thousand. That would be fine for nearly all practical applications: plumbing, working out how much fertiliser you need for a sports field, fitting hoses onto pipes, etc.

But consider the Voyager 1 spacecraft – the most distant object we humans have ever sent into space. It's about 20 billion kilometres away – about four times the distance of Pluto from us. To know its position accurate to three centimetres, we need to know π to only 15 places. (That's surprisingly few – my first guesstimate was 100 places.)

And consider the biggest thing we know – the Entire Known Universe. It has a radius of about 46 billion light years. If we want to find out the circumference of the Entire Known Universe to within the diameter of a tiny hydrogen atom, we need only 39 or 40 places of π.

@DoctorKarl **What's the best bet for life in the Solar System?**

So when it comes to measuring the size of the Universe, it really is just all π in the Sky.

THROWING STICKS PI

During the American Civil War, Captain O.C. Fox was forced to spend weeks in hospital recovering from a head wound. He passed much of that time throwing strands of fine steel wire onto a wooden board covered with parallel lines. Sometimes the wire landed between the lines, and sometimes on the lines.

After 1100 throws, Captain Fox was able to work out π to two decimal places – 3.14.

How?

Captain Fox used a method called "Buffon's Needle". In 1777, Georges-Louis Leclerc, Comte de Buffon, published a paper that was the basis of Captain Fox's pastime. Buffon considered what would happen if you threw a coin onto a floor that was covered with square tiles. At the time, this activity was very popular in his social circles.

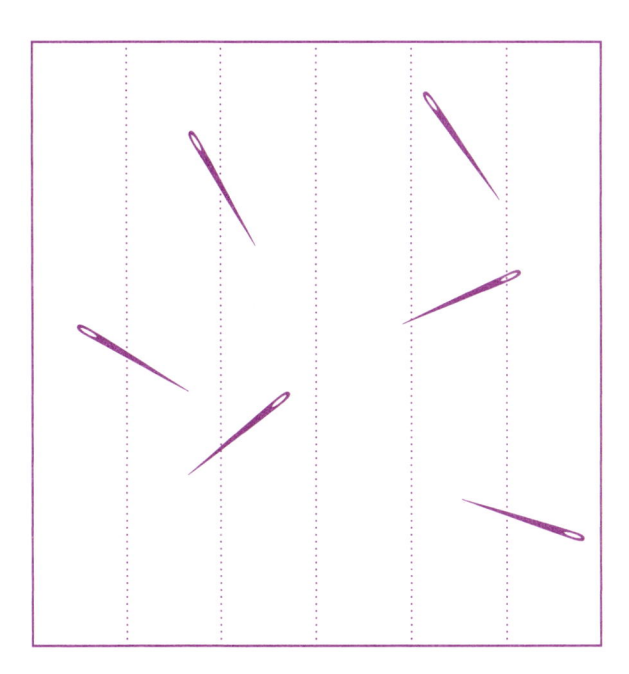

Life in Solar System? I would have Enceladus up with Europa (both have oceans of liquid water covered with ice).

People would place bets on whether the coin would land on a line or not. But what was the probability of each event?

Buffon became the first person to bring calculus into Probability Theory. He started with a sewing needle, not a coin, and worked out the basic maths that Captain Fox later used. The probability of a needle (or a stick, or a steel wire) landing on a line is $\frac{2}{\pi} \times \frac{1}{d}$ where "l" is the needle length, and "d" is the distance apart of the parallel lines. If the length of the needle equals the distance between the parallel lines, the probability of a needle landing on a line is $\frac{2}{\pi}$

By solving the equation using his experimental data, Captain Fox managed to work out π to two decimal places. If he had wanted to get that third decimal place of π, he would have needed around 1000 years or about 27,500,000 throws. This method of getting π works – but getting the data takes a very long time.

SHOTGUN PI

You can also work out π with a 28-inch barrel Mossberg 500 Pump-Action Shotgun (although any well-made shotgun should work).

Why use a shotgun?

Well, as the authors, Vincent Dumoulin and Félix Thouin from the University of Montreal, point out, once the Zombie Apocalypse comes, society will temporarily lose a lot of essential scientific information. Therefore, we need a "robust, yet easily applicable method to estimate Pi", especially "using everyday tools". And sure, come the Zombie Apocalypse, what would be more of an "everyday tool" than a Pump-Action Shotgun?

A rifle fires a single bullet, but a shotgun fires many small particles, called "shot". The shot pellets spread as they leave the barrel. They generally land within a circle, and are more concentrated toward the centre of that circle. Depending on your cartridge, and the "choke" on the barrel, a typical spread at 8 metres might be about 20 centimetres.

So where does our beloved constant, π, which seems to pop up everywhere in the Universe, appear in this situation?

Start with a square as your target. Draw a quarter circle from one corner to the opposite corner.

@DoctorKarl Why are you against nuclear power? Nuclear weapons don't equal nuclear power.

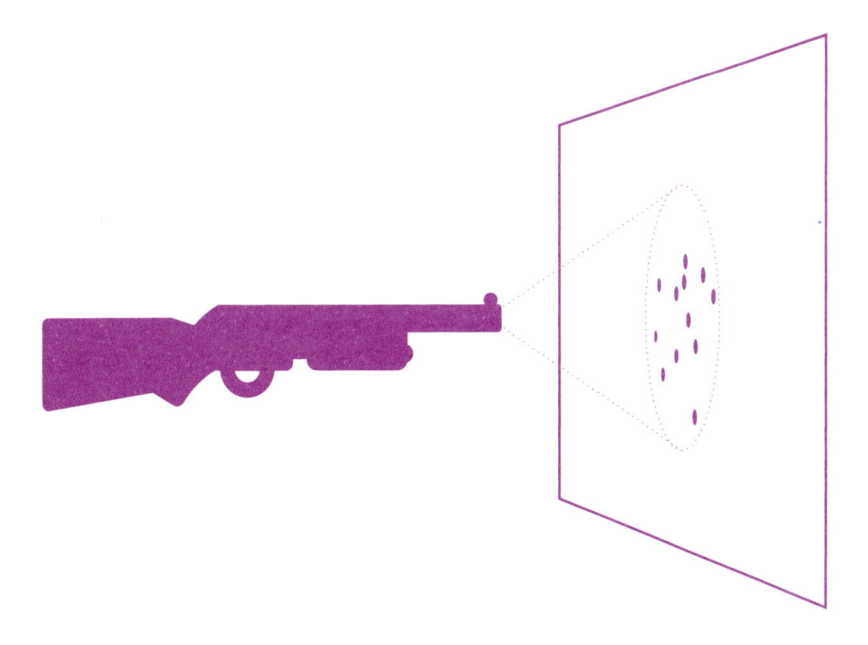

If you then drop a large number of small particles (such as sand, rice, or yes, lead shot) randomly on this target, the ratio that land within the circle should be about $\pi/4$, or roughly 79 per cent.

According to the authors, after the Zombie Apocalypse, there won't be much rice around, but lots of shotguns. So, in the interests of "historical accuracy", the authors then loaded their shotgun with cartridges composed of 3 dram equivalent of powder and 32 grams of #8 lead pellets. They then fired the cartridges at a target 20 metres away, with an average muzzle velocity of 1200 kilometres per hour.

Unfortunately, in real life, shotgun pellets do not land randomly inside their target circle. They are concentrated towards the middle, following a normal Bell Curve distribution. Other factors that work against pure randomness include orientation of the shotgun to the target (whether it's square-on, or side-on), height of the shooter, wind direction, and so on.

So Vincent Dumoulin and Félix Thouin used some fancy mathematics called "Importance Sampling". This would bring the locations of where the shotgun pellets landed back to random. They fired 200 shots at the target, which gave them 30,857 holes. (That's about 150 pieces of lead shot per cartridge.) They then picked a random subset of 10,000 holes, and applied their fancy mathematics. They needed to know how far from random these 10,000 holes were. This gave them an equation, which they

Several countries have used "peaceful" nuclear reactors to make nuclear weapons.

PI IS RANDOM: OR IS IT?

In his novel *Contact*, Carl Sagan has his characters find a pattern in π, following contact with aliens from Vega about 25 light years away.

Another set of aliens had left messages inside certain numbers. When the heroine examines π in Base 11, she eventually finds a circular pattern. This is supposedly proof of a higher intelligence in the Universe.

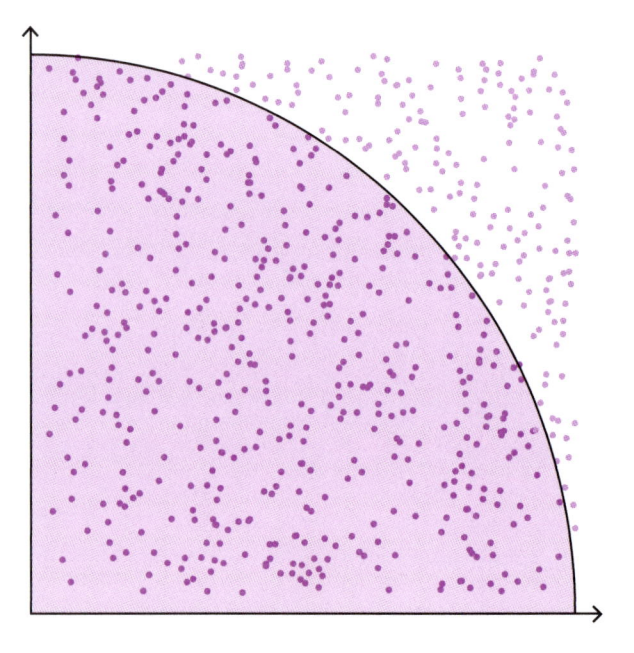

applied to "shift" the remaining 20,857 holes in their battered target – to make them closer to a random scattering.

And then they worked out what ratio of the holes were inside the quarter circle. From this, they worked out π.

How close were they? Instead of 3.142, they got 3.131 – within about 0.33 per cent of the true value. They conclude that "should a tremendous civilisation collapse occur", their "ballistic . . . (random) methods . . . constitute reliable ways of computing mathematical constants".

ANOTHER WAY TO WORK OUT PI

The following series

$$1 + \tfrac{1}{4} + \tfrac{1}{9} + \tfrac{1}{16} + \tfrac{1}{25} + \tfrac{1}{36} \ldots = \pi^2/6$$

... presented by Leonhard Euler to the St Petersburg Academy of Science in 1735 as the solution to the famed "Basel Problem", is yet another way to get π.

HUMAN COMPUTER OF PI

One of the most powerful "human computers" of the 19th century was Johann Martin Zacharias Dase, from Hamburg in Germany. He suffered from epilepsy all his life.

He made his living on the carnival circuit in Germany, Denmark and England by multiplying large numbers in his head for crowds. He never used paper – it was all purely mental. He earned extra money by hiring himself out to mathematicians who needed some arithmetic done.

He once multiplied 79,532,853 by 93,758,479, and got the right answer in 54 seconds. The square root of a hundred-digit number took longer – about 52 minutes. It took even longer (8 hours and 45 minutes) to multiply a pair of hundred-digit numbers.

Was that his limit of his mental arithmetic ability? Not on your life. He could run a single calculation, in his head, over weeks!

He would stop at bedtime, store his calculations in his mind overnight while getting a restful sleep, and then restart in the morning. But he never got tired while calculating.

Occasionally, he would have the equivalent of a modern computer System Crash – sometimes due to a headache. In the case, he would have to start from the beginning.

In 1844, the mathematician L.K. Schulz von Strassnitsky employed him to work out π. Dase calculated π to 200 decimal places – entirely in his head. It took him two months of solid thinking. At that time, nobody else had worked out π to that many digits – it was a world record.

Cure? No. Prevention? Maybe. Main factor? Lack of bright light when reading → change in dopamine levels → eyeball grows bigger → myopia?

PI IS RANDOM: FUN SEQUENCES

Around the 300,000,000th digit of Pi, these eight digits appear – 88 888 888.

A bit further, ten sixes pop up – 6 666 666 666.

And a bit past the 500,000,000th digit, there's this cute sequence – 123 456 789.

PI IS RANDOM: ZERO COINCIDENCES

With fractions, we can predict what the value of any particular digit (10th, 100th, etc) will be.

We cannot predict any digit of π simply by looking at the pattern of digits before it – we have to work it out.

For example, $\frac{1}{7}$ repeats itself. It's equal to 0.142 857 142 857 142 857 etc. But π never repeats itself – at least, not up to around 13 trillion digits, as of late 2015.

But there's something odd with the quadrillionth digit (i.e., a thousand trillion) and the two quadrillionth digit. They are both equal to zero. (Nowadays we can, with massive computing power and lots of time and electricity, work out any digit in Pi, without having to first work out the digits before it.)

If the three quadrillionth digit of Pi turns out to be zero, I'm going to begin to think about suspecting a pattern.

It doesn't mean anything – it's just random statistical noise, or as most people would say, a coincidence.

THE BIBLE SAYS PI = 3?

In the Old Testament, 1 Kings 7:23 (if only it had been 7:22) says, "And he made a molten sea [a bowl], ten cubits from one brim to the other: it was round all about . . . and a line of 30 cubits did compass it round about."

A cubit is the distance from your elbow to the tip of your outstretched finger, or about 45 centimetres. This was a pretty big bowl to have cast from brass – nearly five metres across. If you divide the circumference (30 cubits) by the diameter (10 cubits), you get a value for π of 3. That's wrong by about 3 per cent.

But what if we factor in the thickness of the wall of the bowl? Suppose you measured the diameter on the inside, but measured the circumference on the outside? (The quote says "did compass it round about", so we have to measure the outside circumference.) Assume the thickness of the wall was about 10 centimetres. Then π drops even more, to around 2.9. That's even more wrong.

I guess the author of 1 Kings 7:23 was more interested in the broad sweep of the story, rather than the pedantic mathematics.

Problems with nuclear 1) potential to make nuclear weapons 2) waste (95% of energy remains in waste) 3) Chernobyl.

INDIANA SAYS PI = 3.2?

This story begins in the early 1890s with a physician, Edward J. Goodwin, from Solitude, Indiana. He was also an amateur mathematician. He was wrongly convinced that he had solved an ancient problem called "Squaring the Circle" – even though a professional mathematician, Ferdinand von Lindemann, had already proved this was impossible about a decade earlier, in 1882.

Squaring the Circle means constructing a square that has the same area as a circle – but using ONLY a compass and a straight edge. Dr Goodwin didn't care that it was impossible to Square the Circle. He came up with an incorrect "proof". One part of his proof stated "the ratio of the diameter and circumference is as five-fourths to four". Ignoring a few thousands years of work by real mathematicians, in effect Goodwin had declared that π equalled 3.2!

In fact, he went on to write that having π equal to 3.14 is "wholly wanting and misleading in its practical applications". For good measure, Goodwin then "solved" a few more impossible-to-solve problems such as Doubling the Cube and Trisecting the Angle.

You think that all of this would have alerted people he was on the wrong track – but no. He forced a journal, American Mathematical Monthly, to publish his supposed "proof". In 1894, they released it, but with the disclaimer "published by request of the author".

Goodwin went one step further – he then copyrighted his incorrect "proof". His aim was to collect royalties from mathematicians and businesses that used his incorrect method. But being a good Citizen of Indiana, he didn't want the local schoolchildren to be deprived of his brilliance. So he generously offered his incorrect masterpiece for free to the State of Indiana – if they would crazily pass a bill that said that his proof was now part of the Law of Indiana.

And so, on 6 February 1897, the House of Representatives of the State of Indiana unanimously passed Bill 246 introducing a New Mathematical Truth. In fact, the Bill had the rather long title of:

"A Bill for an act introducing a new mathematical truth and offered as a contribution to education to be used only by the State of Indiana free of cost by paying any royalties whatever on the same, provided it is accepted and adopted by the official action of the Legislature of 1897."

@DoctorKarl IntraVenous coconut water to replace fluid loss? True or False?

The Bill also incorporated the following praise for the "modest" Goodwin: "be it remembered that these noted problems had long since been given up by scientific bodies as insolvable mysteries and above man's ability to comprehend."

By a fortunate coincidence, a mathematician, Professor C.A. Waldo of Purdue University, happened to be at the Indiana statehouse. He was there to make sure that the Indiana Academy of Science received its annual appropriation. He was invited to be introduced to Dr Goodwin, but refused, saying that he had already met enough crazy people. He was able to get the ear of some senators. As a result, the Senate did not adopt Bill 246, and π remained as it had always been – which was fortunate, as π is a mathematical constant.

Incorrect. Often claimed, but not backed up in Medical Literature (I made this mistake myself, previously).

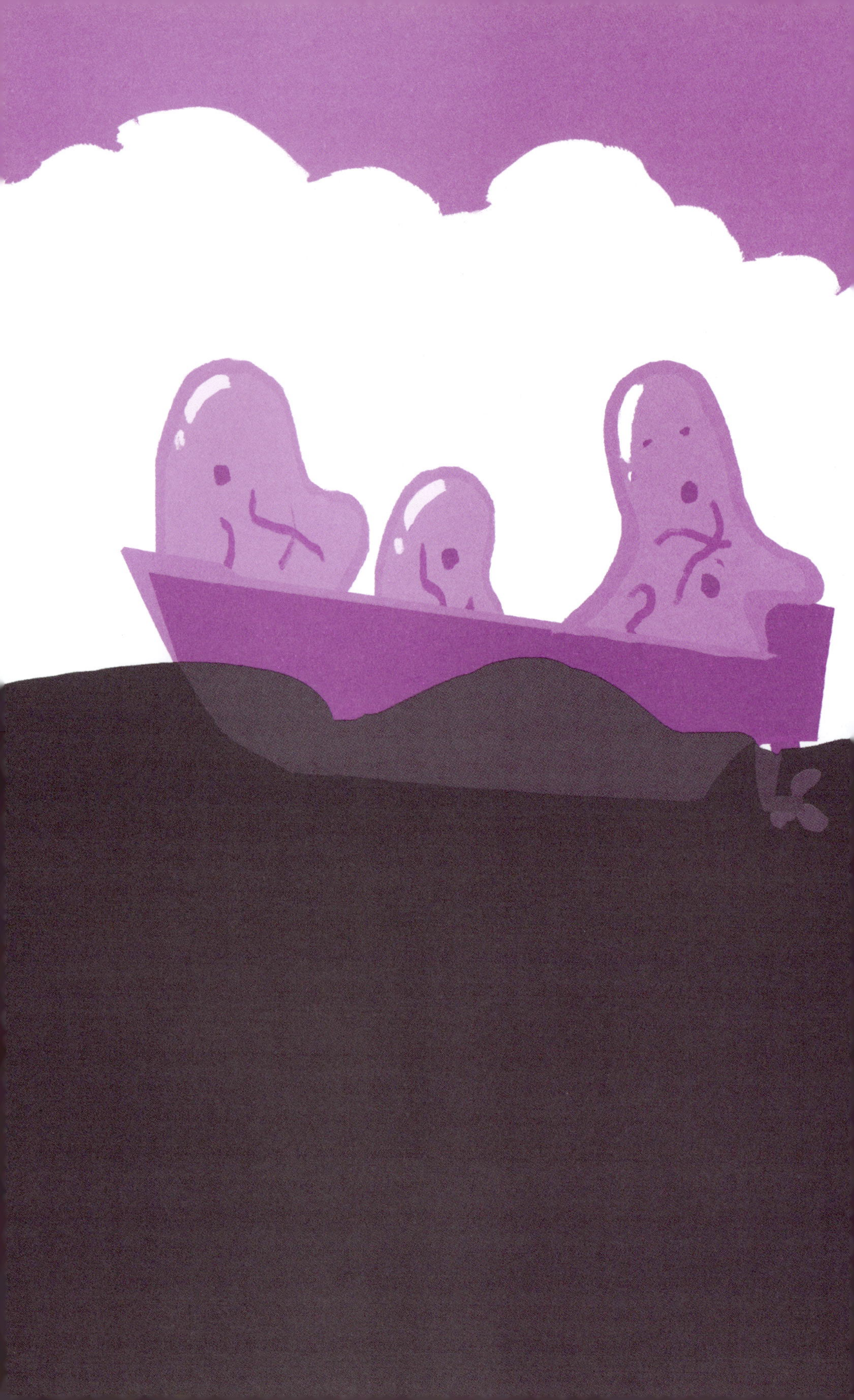

39

ELECTRIC MOTORS IN BACTERIA

BACTERIA ARE EVERYWHERE – THEY MAKE UP HALF THE CELLS IN YOUR BODY.

You don't normally think of bacteria as being technologically State-Of-The-Art. So, it's amazing that some bacteria use propellers (powered by tiny electric motors) to swim in their environment. Even more amazingly, these electric motors literally throw themselves together. Finally, they are much more efficient than any motor we can currently build.

BACTERIA 101

Back in 1676, Antony van Leeuwenhoek, the famous Dutch microscope maker, described what he saw when he examined a bacterium through one of his excellent microscopes: "its belly is flat and provided with [a] divers incredibly thin feet, or little legs, which moved very nimbly". Those "little legs" are called "cilia" (if short) or "flagella" (if longer). They are made mostly from proteins. In general, to work they need to be in at least a thin film of liquid. So bacteria can swim in the ocean by moving their "little legs", and human cells in your airways can use cilia to push dust upwards and out.

For centuries, everybody thought that cilia and flagella just waved back and forth – like the tail of a fish. But in the mid-1970s, we proved that some of them rotated like little propellers. And more recently, by studying the bacterium *E. coli*, we have begun to understand how these organic motors assemble themselves.

E. coli is shaped like a stubby rod – about 2 microns long (that's 2000 nanometres, or billionths of a metre) and 1 micron in diameter. (A micron can also be written as "μ", and is one millionth of a metre.) *E. coli* reproduces very quickly – every 20 minutes. The cell wall is a distinct physical structure that keeps the outside out, and the inside in. This cell wall is about 7 to 8 nanometres (billionths of a metre) thick, and is made of various chemicals – sugars, proteins, fats.

Each *E. coli* bacterium has about four or five "wavy arms" randomly distributed on its surface. In the past, we used to call them "flagella", but now we call them "flagellar filaments". Each filament is spun by its own astonishing organic electric micro-motor.

BACTERIAL FLAGELLAR MOTOR

Even though this Bacterial Flagellar Motor is tiny, it's made of three even smaller parts – the electric motor buried in the cell wall, the flagellar filament, which is about 10 microns long, and the "hook", which is a flexible joint joining the motor and the filament.

A small part of the electric motor is outside the body of the *E. coli*, another small part is inside the body of the *E. coli*, while a significant part of the electric motor actually sits within (and across) the cell wall.

The electric motor is about 45 nanometres in diameter – this is truly tiny. It can rotate at up to 50,000 revolutions per minute – about three times faster than the engine in a Formula One racing car. It's made from about 20 proteins, with a total molecular mass of 11 Megadaltons. (Hydrogen is about 1 dalton, water is 18 daltons, insulin is about 5000 daltons. So the Bacterial Flagellar Motor is about 2000 times heavier than a single insulin molecule.)

It's an Electrochemical Motor – it gets its energy supply from charged ions crossing the cell membrane. (The energy efficiency is close to 100 per cent – much higher than we can achieve with our human-made electric motors.)

Just like a human-made electric motor, this Bacterial Flagellar Motor has a rotor that spins. Again, like a human-made electric motor, it has "stators". (In electric motors, a current through the stator causes the rotor to spin.) *E. coli* has up to 11 stators – the more stators, the more powerful is the motor.

Unlike the electric motors that we humans make, this dynamic microscopic molecular machine is constantly being re-built and re-configured on the run – the stators continuously come and go. In fact, the average "lifetime" of any given stator in the Bacterial Flagellar Motor is of the order of half a minute.

Too heavy + Gravity. Earth loses ~90,000 tonnes/year of H & He. Gains 50,000 tonnes/year of meteorites → loss ~40,000 tonnes/year.

MOVING BACTERIUM

Cleverly, the bacterial flagellar filaments behave very differently when they rotate clockwise or anticlockwise.

When they rotate anticlockwise, they automatically wind themselves into spinning bundles that push the bacterium forward through its liquid environment. The bacterium can now swim at up to 100 microns per second (or about 50 to 100 times its own body length per second).

But when the filaments rotate clockwise, they unwind and splay out from the bundle. Suddenly the bacterium goes into a tumble, and ends up pointing in a different direction. This lets it change direction.

It turns out that there are many different versions of this Bacterial Flagellar Motor. Evolution has taken many separate pathways to arrive at different motors that are optimised for different bacteria and their different environments.

For example, the bacterium *Campylobacter jejuni* lives in a high-nutrient, high-viscosity environment – your gut. Its motor is optimised to be so powerful that it can easily push through the gut wall – and cause food poisoning. (It does this by having a bigger rotor, which means a bigger "moment arm", which means more torque.)

On the other hand, *Caulobacter crescentus* lives in low-nutrient, low-viscosity freshwater environments, and its motor is optimised for speed.

The Bacterial Flagellar Motor is the "pinnacle of evolutionary bionanotechnology: a self-assembling nanoscale electric rotary motor that performs at a higher speed and with greater efficiency than any man-made device".

@DoctorKarl **Does Mars have a north pole?**

SELF-ASSEMBLY!

"Self-Assembly" is exactly what I mean. This is not a misprint, nor is it a mistake. These electric motors actually assemble themselves.

Once the components have been manufactured, they come together by themselves. It's almost as though you have a bag full of random LEGO pieces, and after you give it a vigorous shake, suddenly the pieces assemble themselves into a perfectly constructed *Star Wars* TIE fighter, or a perfect Pirate Ship.

We're still learning how the Bacterial Flagellar Motor does this – but we do know that it starts with bits of the rotor. Here, one of the individual components has a "wound-up spring" inside. Only when it matches up perfectly with another identical component do the two "springs" unwind and lock themselves together.

But how do the motors know *where* to assemble? After all, you want your electric motors straddling and punching *through* only the cell wall – and not anywhere else inside the bacterium. So far, we are pretty sure that part of the cell wall (the scaffold upon which the rest of the motor is built) is important. We think that once enough of the components congregate inside the cell wall, they will self-assemble. We also think that once the first stage of self-assembly starts, these initial components and the cell wall provide a platform on which the rest of the motor can be built.

So next time you're struggling to build a LEGO Pirate Ship, try to think like a bacterium. After all, that's where the world's fastest and most efficient nanomachine builds itself, in the dark, with no instructions.

Apparently. Very weak magnetic field of Mars → (solar wind from Sun ripping away atmosphere of Mars). Maybe weak N?

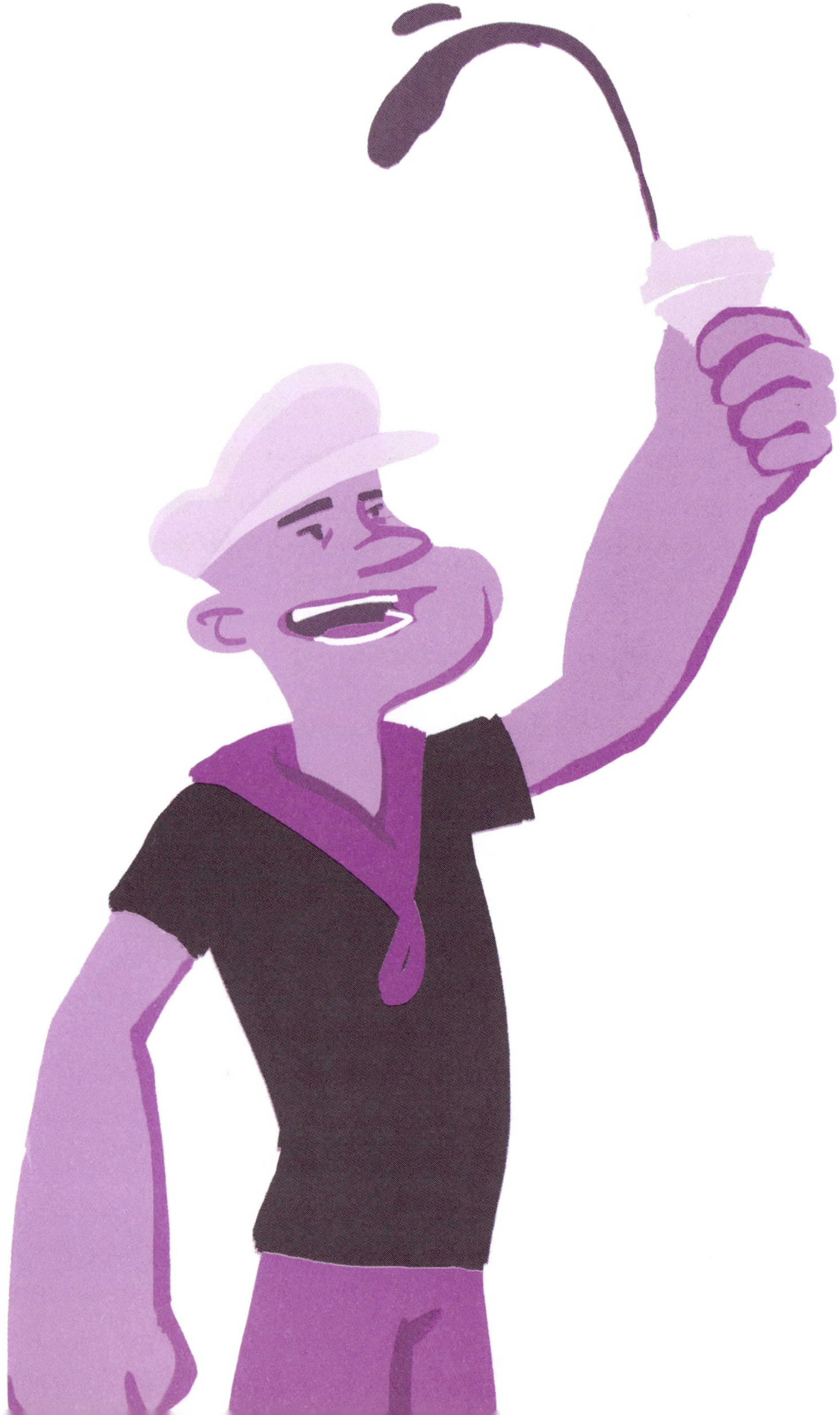

40

COFFEE IS NOW GOOD

COFFEE IS THE WORLD'S MOST POPULAR LEGAL DRUG. IT IS A DRUG, AND SO WE SHOULD REMEMBER THE WORDS OF THE FATHER OF TOXICOLOGY, PARACELSUS: "ALL DRUGS ARE POISONS, WHAT MATTERS IS THE DOSE."

Based on the fact that coffee has been used widely (and overwhelmingly safely) for over a millennium, we should expect that its bad side effects would be fairly minimal – so long as we don't drink too much.

CORRELATION IS NOT CAUSATION: PART 1

"Correlation" is a mutual link between two things. "Causation" is where one thing causes another.

Just for fun, look at the correlation between "per capita margarine consumption in the USA" and "the divorce rate in the state of Maine".

Is the correlation excellent over the years? Yes.

In a surprising (and lucky) break for coffee-drinkers, there is a gradually increasing Ocean of Statistics showing that some of the side effects of coffee actually seem to be good. And interestingly, many of these good side effects appear to have nothing to do with caffeine – they seem to be related to other chemicals in coffee.

Taken with a grain of salt (which itself may or may not be good for your health), the main positive effects seem to be on life expectancy, liver disease, Type II Diabetes, prostate, skin and oral cancers, and heart conditions. However, coffee sometimes has bad outcomes in the case of lung cancer.

STATISTICS 101

But there's a niggling problem – the original Statistics that gave us these findings. Most of the data on these good effects from coffee drinking have popped up "incidentally". (In other words, the studies were not specifically set up to look at coffee drinking and its effects on health.)

This is because most of the data comes from *observational* studies. These studies were set up to look at the actions, activities and health outcomes of huge numbers of people – and over many years, often decades. These studies weren't looking for anything specific – instead, they were just observing (or fishing), to see what would pop up as incidental findings. They were looking for "correlations".

@DoctorKarl How come Evolution hasn't saved us from biting our tongue?

> Does margarine cause divorce?
> No. There is no "causation".
>
> This is what Statisticians mean by the phrase, "Correlation is not causation."

The results on coffee came as secondary data from studies that were looking at other issues. This can mean that the data may not be as statistically robust as we would like. It has a large anecdotal component.

META-ANALYSIS

A meta-analysis tries to be a statistical version of the Wisdom of the Commons. (This is where, in some cases, the "many" can be smarter than the "few".)

A meta-analysis (or meta-study) combines a whole bunch of separate studies that are all looking in the same area. Ideally, if all the individual studies are done to the same high standard, the final meta-analysis should be able to arrive at better results – that is, provide more precise estimates of the result, and with lower uncertainty.

The statistician Karl Pearson published one of the very first meta-analyses in 1904. His meta-analysis combined the data from several smaller clinical studies of typhoid vaccination.

But sometimes, some of the individual studies that make up the meta-study might be flawed. Unfortunately, it's not always possible to see the problems (and sources of bias) in a study just by reading the paper related to that study. If some of the studies are flawed, then the final result of the meta-study is suspect. After all, a result is only as good as the data it's built on.

The old adage still holds: "GIGO", or "Garbage In, Garbage Out".

---

1) Clearances in mouth b/w cheeks & tongue are in mm, with both moving
2) We have evolved to not bite tongue, very rare.

CORRELATION IS NOT CAUSATION: PART 2

A very interesting correlation-causation result popped up from the original Nurses' Health Study.

This observational study was set up in 1976 looked at various factors that might possibly be related to health. The subjects were more than 120,000 female registered nurses in the USA. By 1990, some 14 years later, a very clear correlation had popped up.

LIFE EXPECTANCY

It seems that coffee drinkers live longer.

One meta-study reviewed 20 other studies that included over 970,000 people, while another looked at 17 studies that had over a million participants. They compared those who drank the most coffee with those who drank the least. The heavy coffee drinkers had a 14 per cent lower risk of dying prematurely from any cause. Even having just one or two cups each day dropped the risk of premature death by 8 per cent.

And drinking decaffeinated coffee gave the same advantage. Drinking 2 to 4 cups of decaf per day still kept the risk of premature death at 14 per cent lower.

LIVER

For people with any liver disease, coffee slows down the progression to cirrhosis. Even if someone already has cirrhosis, coffee is linked to lower risks of both death and developing liver cancer.

For a person with Hepatitis C, coffee is linked to better responses to antiviral treatment.

If someone has Non-Alcoholic Fatty Liver Disease, coffee drinking is associated with better outcomes.

Coffee also helps with liver cancer. Liver cancer is the sixth most

> Any nurses who were undertaking Combined Hormone Replacement (CHR) were having fewer cases of Coronary Heart Disease. The correlation seemed obvious – if women took CHR, they were definitely protected from Coronary Heart Disease.
>
> However, while this correlation superficially seemed "real", the causation was not. In fact, taking other factors into account, the correlation went away.
>
> You see, the women who took CHR were wealthier, and also took more care of their health, such as eating well and doing exercise. Once these factors were accounted for, it turned out that CHR had no effect on Coronary Heart Disease – or maybe a small increase.

common cancer in the world, and more frequent in men than in women. According to a meta-study based on six other studies, coffee drinkers dropped their relative risk of liver cancer by 14 per cent for every extra cup of coffee they had each day.

In the case of liver cancer, two natural chemicals in coffee (kahweol and cafestrol) have direct cancer protection and anti-inflammatory properties. They could be causing this good result. In terms of protection from cancer, Collins says that they seemed to "upregulate biochemical pathways in the liver that protect the body from toxins, including aflatoxin and other carcinogenic compounds".

TYPE II DIABETES

You probably know that Type II Diabetes is developing into a major problem around the world. Does coffee make a difference?

Yep, you guessed it, coffee drinkers have a reduced risk of Type II Diabetes. This finding comes from a meta-study of 28 other studies covering over 1 million adults. People who drank three or more cups of coffee each day had their relative risk of developing Type II Diabetes lowered by 21 per cent – as compared to those who never, or only rarely, drank it. Furthermore, if they drank six or more cups each day, the risk was lowered even further – by 33 per cent.

As with Life Expectancy, decaffeinated coffee was also protective

---

Originally → pregnant women → prevent Cretinism in new-born babies.
https://en.wikipedia.org/wiki/Iodine#Biological_role
Also blocks radioactive iodine.

– but not as much as caffeinated coffee. The protective effect was about one third lower than for caffeinated coffee.

It's thought that there are two active ingredients driving this benefit. They are chlorogenic acid, which can improve glucose metabolism and insulin sensitivity, and caffeic acid, which increases how quickly muscles use up the glucose in blood, but which also has both immune-stimulating and anti-inflammatory properties.

PROSTATE CANCER

In American men, prostate cancer is the most frequently diagnosed form of cancer. It's also the second leading cause of cancer death. Happily for coffee drinkers, the links between prostate cancer and coffee are positive.

A meta-study looked at 13 other studies that in total included more than 530,000 men. For every extra 2 cups of coffee that were drunk each day, the risk of prostate cancer dropped by 2.5 per cent. Overall, those who drank the most coffee had a 10 per cent lower risk of getting prostate cancer as compared to those who drank the least. Other meta-studies showed similar results.

People still got improvements in health outcomes with decaffeinated coffee. This implies that the improvements were due to other chemicals in the coffee.

SKIN AND ORAL CANCER

It seems that coffee can reduce skin cancer – at least, in mice. It turns out that some damaged cells can turn into skin cancers. Coffee seems to kill off these damaged cells. As always with animal studies, this research needs to be followed up with human studies. (And I'm still going to keep using sunscreen, protective clothing, a big hat and , to be sardonic, wraparound "high-fashion" sunglasses.)

Another study showed that the risk of oral cancer could be reduced by drinking caffeinated coffee. The protection did not hold for decaffeinated coffee, nor for tea. This finding popped up from the Cancer Prevention Study II, a prospective US cohort study that began in 1982 with testing 968,432 men and women. If they drank more than 4 cups of caffeinated coffee each day, overall they had a 49 per cent lower risk of oral/pharyngeal cancer.

THE HEART

Coffee seems to have a strange J-curve relationship with Heart Failure. It is most protective at four cups per day – leading to an 11 per cent lower risk. At coffee intakes both lower and higher than four cups per day, the risk of Heart Failure rises.

Again, this was not a study that had been specially designed to look at Heart Failure. No, it was a meta-study of five other studies, covering a time period from 1966 to 2011, with 140,220 participants – of whom 6522 had heart failure.

Another meta-study combined 36 studies that involved over 1,270,000 participants. Again, those who had about three to five cups of coffee per day were at the lowest risk of Cardiovascular Disease.

On the bad side, Collins says caffeine (not in decaffeinated coffee) can "increase blood pressure in the short term and plasma homocysteine, another heart disease risk factor".

LUNG CANCER

Now for the bad news – at least for caffeine.

Meta-studies of over 100,000 adults found that those who drank the most coffee had a 27 per cent higher relative risk of getting lung cancer.

But included in this meta-study were two studies that looked at decaffeinated coffee – and which gave the opposite result. People who had lots of decaffeinated coffee had a 34 per cent lower relative risk for lung cancer.

Spinal discs (= nucleus pulposus + disc annularis) shrink, & there's some shrinkage/osteoposoris of vertebral bodies.

PREGNANCY

It's good to be cautious in pregnancy.

But it seems safe to have a coffee or two each day – even in pregnancy.

There was previously thought to be a link between coffee and miscarriage, but it is now somewhat disputed. Consider severe morning sickness. It's often a sign of good implantation of the embryo into the uterus – and ultimately, a healthy, strong baby. But it's also related to nausea. So a woman with severe morning sickness might specifically avoid coffee because she felt too sick to drink it – and would be more likely to deliver a healthy baby.

Result? A casual look at the data would show that women who avoided coffee would be more likely to have a healthy baby, but on further examination this might be a case of correlation rather than causation.

COFFEE: GOOD OR BAD

A good study does not cherry-pick the data to get the desired result. (Cherry-picking data is what Big Tobacco and the denialists of Global Warming have been doing for decades.)

On one hand, there is a growing body of evidence that coffee might have good effects on your health. And decaffeinated coffee also seems to be a good option in most cases.

On the other hand, practically all of these studies are meta-studies – collections of smaller studies, of variable quality. Furthermore, these smaller studies were mostly designed to follow a large group of people for one or more decades, and see what popped up.

What we really need are randomised controlled trials, which would be specifically set up in advance to explore and tease out the effects of coffee.

So bottoms up, if coffee is your poison. At this stage, I won't feel too bad about having coffee. But I'm not voting to add it to the drinking water just yet...

COFFEE AND CALORIES

Coffee can be loaded with energy – and I don't mean "caffeine", I mean calories or kilojoules.

A black coffee has about 20 kilojoules – and zero fat and zero carbohydrates. So if you want to have coffee, it's not "fattening".

The "average" adult daily requirement for kilojoules is about 8700 per day. (Go to www.8700.com.au and plug in your information to get a number more specific to you.)

A McDonald's large mocha (10.2 grams of fat, 31.8 grams of carbohydrates) has about 1100 kilojoules.

In America, the Cold Stone Creamery Gotta-Have-It-sized Lotta Caramel Latte takes the cake. It has 90 grams of fat (about half to a third of a block of butter!), 223 grams of carbs – and 7500 kilojoules. That is practically your total energy needs for an entire day. At the same time, it gives you hardly any fibre, minerals, vitamins or protein.

Cold water on skin → peripheral vasoconstriction → more blood to kidneys → ↑Glomerular Filtration Rate → more urine → wee more.

41 WEIGHT LOSS VIA EXERCISE

If your goal is to lose weight, the best time to exercise is first thing in the morning on an empty stomach.

At least, that's according to a 2010 study in Belgium. The researchers looked at some healthy young men – average age 21, with an average body weight of 71 kilograms. They persuaded them to increase their daily kilojoule intake by 30 per cent every day for six weeks. The new diet was very rich in fat. In fact, fat made up 50 per cent of their total intake of kilojoules.

The researchers split the men into three groups. The first group did no extra exercise. They gained 3 kilograms over the six weeks. Their blood chemistry showed that they were developing insulin resistance. Their muscles were loaded with fat cells.

The second group did endurance exercise training (four days per week) in the mid-morning – after breakfast. They gained about 1.4 kilograms. Their blood chemistry was a little worrying, but certainly a lot better than the group that did no exercise.

The third group did the same exercise program, but on an empty tummy and "in a fasted state" – i.e. before breakfast. They gained only 0.7 kilograms, and their blood chemistry was virtually normal.

We know that skeletal muscle "plays a major role in glucose metabolism accounting for ~75 per cent of whole-body insulin-stimulated glucose uptake". But it's interesting that the timing of muscle activity had an effect.

There were some limitations to this study. For example, the sample size was very small – only 27 volunteers in all.

Insulin resistance is increasing in Western societies. We know that eating less, combined with more physical activity, can help with insulin resistance. But many of us do not have enough spare time to fit more exercise in.

So if any of this interests you, try waking up earlier and exercising on an empty stomach.

**Fire often relatively complete combustion → invisible CO_2, H_2O, NO_x etc.
Smouldering is incomplete combustion → buncha visible chemicals.**

REFERENCES

01 ALCOHOL AND HEARING

"Frequency Selective Effects of Alcohol on Auditory Detection and Frequency Discrimination Thresholds", by P. Pearson et al., *Alcohol and Alcoholism*, September–October 1999, Vol. 34, No. 5, pages 741–749.

"Acute Effects of Alcohol on Auditory Thresholds and Distortion Product Otoacoustic Emissions in Humans", by Juen-Haur Hwang et al., *Acta Oto-laryngologica*, October 2003, Vol. 123, No. 8, pages 936–940.

"Auditory Assessment of Alcoholics in Abstinence", by Sandra Beatriz Ribeiro et al., *Brazilian Journal of Otorhinolaryngology*, July–August 2007, Vol. 73, No. 4, pages 452–462.

"The Acute Effects of Alcohol on Auditory Thresholds", by Tahwinder Upile et al., *BMC Ear, Nose and Throat Disorders*, 18 September 2007, Vol. 7, No. 4, http://bmcearnosethroatdisord.biomedcentral.com/articles/10.1186/1472–6815–7-4

02 PASTEURISED MILK GOES OFF?

"Milk That Lasts", by C. Claiborne Ray, *The New York Times*, 6 December 2005.

"Got Milk? Don't Get Raw Milk! A Cautionary Tale", by Robert V. Tauxe, *CDC Expert Commentary*, 16 May 2011, http://www.medscape.com/viewarticle/742147

"Nonpasteurized Dairy Products, Disease Outbreaks, and State Laws – United States, 1993–2006", by Adam J. Langer et al., *Emerging Infectious Diseases*, March 2012, Vol. 18, No. 3, pages 385–391.

"Raw Deal: California Cracks Down on an Underground Gourmet Club", by Dana Goodyear, *The New Yorker*, 30 April 2012.

"Raw Milk Questions and Answers", The US Centers for Disease Control and Prevention, 20 February 2015, http://www.cdc.gov/foodsafety/rawmilk/raw-milk-questions-and-answers.html#related-outbreaks

"Got 'Raw' Milk?", by Rebecca Simpson, Sydney Environment Institute, 8 September 2015, http://sydney.edu.au/environment-institute/news/got-raw-milk/

03 DOGS TILT HEAD

"Why Do Some Dogs Tilt Their Heads When We Talk to Them?", by Stanley Coren, 11 December 2013, *Psychology Today Canine Corner* blog, https://www.psychologytoday.com/blog/canine-corner/201312/why-do-some-dogs-tilt-their-heads-when-we-talk-them

04 VAMPIRE BLOOD DRINKING

"Vampire Label Unfair to Porphyria Sufferers", by Claus A. Peierach, *The New York Times*, 13 June 1985.

"Did Vampires Suffer from the Disease Porphyria – Or Not?", by Cecil Adams, *The Straight Dope*, 7 May 1999, http://www.straightdope.com/columns/read/1321/did-vampires-suffer-from-the-disease-porphyria-or-not

"'Vampires' Strike Malawi Villages", by Raphael Tenthani, *BBC News*, 23 December 2002, http://news.bbc.co.uk/2/hi/africa/2602461.stm

"Cinema Fiction Versus Physics Reality: Ghosts, Vampires and Zombies", by Costas J. Efthimiou and Sohang Gandhi, arXiv:physics/0608059v2, 27 August 2007, http://arxiv.org/abs/physics/0608059v2

"Physics Students Calculate How Long A Vampire Needs To Drink Your Blood", by Esther Inglis-Arkell, *Gizmodo*, http://www.gizmodo.com.au/2016/03/physics-students-calculate-how-long-a-vampire-needs-to-drink-your-blood/, 19 March 2016.

05 PERPETUAL PRESENT

"Severely Deficient Autobiographical Memory (SDAM) in Healthy Adults: A New Mnemonic Syndrome", by Daniela J. Palumbo et al., Neuropsychologia, June 2015, Vol. 72, pages 105–118.

"In a Perpetual Present", by Erika Hayasaki, Wired, April 2016.

06 PERPETUAL PAST

"A Case of Unusual Autobiographical Remembering", by Elizabeth S. Parker et al., *Neurocase*, February 2006, Vol. 12, No. 1, pages 35–49.

"A Case of Hyperthymesia: Rethinking the Role of the Amygdala in Autobiographical Memory", by Brandon A. Ally et al., *Neurocase*, 23 April 2012, Vol. 19, No. 2, pages 166–181.

"Behavioural and Neuroanatomical Investigation of Highly Superior Autobiographical Memory (HSAM)", by Aurora K.R. LePort et al., *Neurobiology of Learning and Memory*, July 2012, Vol. 98, No. 1, pages 78–92.

"The Amazing Memory Marvels", by Kayt Sukel, *New Scientist*, 18 August 2012, Vol. 215, No. 2878, pages 34–37.

"False Memories in Highly Superior Autobiographical Memory Individuals", by Lawrence Patihis et al., *PNaS*, 18 November 2013, Vol. 110, No. 52, pages 20947–20952.

"Remembrance of All Things Past", by James L. McGaugh and Aurora K.R. LePort, *Scientific American*, February 2014, Vol. 310, No. 2, pages 40–45.

"Individual Differences and Correlates of Highly Superior Autobiographical Memory", by Lawrence Patihis, *Memory*, 28 August 2015, http://dx.doi.org/10.1080/09658211.2015.1061011

"Highly Superior Autobiographical Memory: Quality and Quantity of Retention Over Time", by Aurora K.R. LePort et al., *Frontiers in Psychology*, Vol. 6, 21 January 2016, PMC4720782.

"A Cognitive Assessment of Highly Superior Autobiographical Memory", by Aurora K.R. LePort, *Memory*, 16 March 2016, http://dx.doi.org/10.1080/09658211.2016.1160126

"Roots of Indelible Memories Traced", by Helen Thomson, *New Scientist*, Vol. 230, No. 3071, 9 April 2016, page 10.

07 HEAT WAVE

"Presenteeism: At Work – But Out of It", by Paul Hemp, *Harvard Business Review*, October 2004, pages 49–58.

"Death Toll Exceeded 70,000 in Europe During the Summer of 2003", by Jean-Marie Robine et al., *Comptes Rendus Biologies*, February 2008, Vol. 331, No. 2, pages 171–178.

"Higher Temperatures Seen Reducing Global Harvests", by Constance Holden, *Science*, 9 January 2009, Vol. 323, No. 5911, page 193.

"Historical Warnings of Future Food Insecurity with Unprecedented Seasonal Heat", by David S. Battisti et al., *Science*, January 9, 2009, pages 240–244.

"NSW Health Urges Caution During Impending Heat Wave", NSW Health, 29 November 2012, http://www.health.nsw.gov.au/campaigns/beat_the_heat/pages/default.aspx

"Historic and Future Increase in the Global Land Area Affected by Monthly Heat Extremes", by Dim Coumou and Alexander Robinson, *Environmental Research Letters*, 14 August 2013, Vol. 8, No. 3, 034018.

"Do You Know the Signs of Heat Stress?", *ABC Health and Wellbeing: The Pulse*, 15 January 2014, http://www.abc.net.au/health/thepulse/stories/2014/01/15/3925505.htm

"Mechanism Behind Mega-Heatwaves Pinpointed: Two Recent Record Hot Spells Traced to Feedback Loop Between Dry Soils and Trapped Air", by Hannah Hoag, *Nature*, 20 April 2014, http://www.nature.com/news/mechanism-behind-mega-heat-waves-pinpointed-1.15078

"Multimodel Assessment of Extreme Annual-Mean Warm Anomalies during 2013 over Regions of Australia and the Western Tropical Pacific", by Thomas R. Knutson et al., in "Explaining Extremes of 2013 from a Climate Perspective", *Bulletin of the American Meteorological Society*, September 2014, Vol. 95, No. 9, S26–S96.

"Heat Is Australia's Number One Natural Killer", by Sara Phillips, *ABC Environment*, 4 September 2014, http://www.abc.net.au/environment/articles/2014/09/04/4081144.htm

"Feeding a Hungry Nation: Climate Change, Food and Farming in Australia", by L. Hughes et al., Climate Council of Australia, Potts Point, 2015.

"Can a Hot Drink Help Keep You Cool in Hot Weather?", by Nicola Garrett, *ABC Health and Wellbeing*, 5 February 2015, http://www.abc.net.au/health/talkinghealth/factbuster/stories/2015/02/05/4173208.htm

"Severe Heat Costs the Australian Economy US$6.2 Billion a Year", by Michael Slezak, *New Scientist*, 4 May 2015.

"Heat Stress Causes Substantial Labour Productivity Loss in Australia", by Kerstin K. Zander et al., *Nature Climate Change*, 4 May 2015, Vol. 5, No. 7, pages 647–651.

"India Heat Wave Kills More than 500 People", by Jason Burke, *The Guardian*, 25 May 2015, http://www.theguardian.com/world/2015/may/25/india-heatwave-deaths-heatstroke-temperatures

"The Climate Context For India's Deadly Heatwave", by Andrea Thompson, *Scientific American*, 4 June 2015, http://www.scientificamerican.com/article/the-climate-context-for-india-s-deadly-heat-wave/

"The Deadly Combination of Heat and Humidity", by Robert Kopp et al., *The New York Times*, 6 June 2015.

"Future Population Exposure to US Heat Extremes", by Bryan Jones et al., *Nature Climate Change*, July 2015, Vol. 5, No. 7, pages 652–655.

"In Karachi, a Fatal Mix of Heat and Piety", by Mohammed Hanif, *The New York Times*, 26 June 2015.

"Extreme Weather Could Trigger Frequent Global Food Shocks", by Michael Le Page, *New Scientist*, 15 August 2015.

"Bushfires, Heatwaves and Early Deaths: The Climate Is Changing Before Our Eyes", by Tim Flannery, *The Guardian*, 26 August 2015.

"Beat The Heat", NSW Health, 16 October 2015, http://www.health.nsw.gov.au/environment/beattheheat/pages/default.aspx

"European Summer Temperatures Since Roman Times", by J. Luterbacher et al., *Environmental Research Letters*, Vol. 11, No. 2, 29 January 2016, 024001.

"Decades-Long Heatwaves May Hit Europe as Climate Change Bites", by Fred Pearce, *New Scientist*, 29 January 2016, https://www.newscientist.com/article/2075571-decades-long-heatwaves-may-hit-europe-as-climate-change-bites/

"Influence Of Extreme Weather Disasters on Global Crop Production", by Corey Lesk et al., *Nature*, 7 January 2016, Vol. 529, No. 7584, pages 84–87.

"Reaping What We Sow", by Irakli Loladze, *New Scientist*, 9 April 2016, Vol. 230, No. 3068, pages 18–19.

08 ROADS OF ICE

"Ice Lubrication for Transporting Heavy Stones to the Forbidden City in 15th–16th Century China", by Jiang Li et al., *40th Leeds-Lyon Symposium on Tribology and Tribochemistry Forum 2013*, 4–6 September 2013, Lyon, France, http://tribo-lyon2013.sciencesconf.org/15396/document

"Ancient Builders Slid 100-Ton Rocks on Ice Paths to Construct China's Forbidden City", by Nidhi Subbaraman, *NBC News*, 4 November 2013, http://www.nbcnews.com/science/ancient-builders-slid-100-ton-rocks-ice-paths-construct-chinas-8C11522440

"Beijing's Forbidden City Built on Ice Roads", by Dan Vergano, *National Geographic*, 5 November 2013, http://news.nationalgeographic.com/news/2013/10/131104-china-ice-road-forbidden-city-culture-science/

"China's Forbidden City Built with Giant 'Sliding Stones'", by Charles Q. Choi, *Live Science,* 4 November 2013, http://www.livescience.com/40924-how-china-forbidden-city-was-built.html

"Forbidden City Builders Chose Ice Sledge over Wheels", by Colin Barras, *New Scientist*, 4 November 2013, https://www.newscientist.com/article/dn24517-forbidden-city-builders-chose-ice-sledge-over-wheels/

"Forbidden City Built from Stones Dragged on Ice", by Sid Perkins, *Scientific American*, 4 November 2013, http://www.scientificamerican.com/article/forbidden-city-built-from-stone-on-ice/

09 REFUEL CAR WITH ENGINE RUNNING?

"Dubai Petrol Station Blaze Likely Caused by Poorly Maintained Car", *The National* (UAE), 4 June 2015, http://www.thenational.ae/uae/dubai-petrol-station-blaze-likely-caused-by-poorly-maintained-car

"Dubai Motorists Reminded of Safety Regulations when Refuelling", by Nadeem Hanif, *The National* (UAE), 5 June 2015, http://www.thenational.ae/uae/transport/dubai-motorists-reminded-of-safety-regulations-when-refuelling

"Car Catches Fire Twice at Dubai Petrol Station", *The National* (UAE), 9 June 2015, http://www.thenational.ae/uae/transport/car-catches-fire-twice-at-dubai-petrol-station

10 GRAVITATIONAL WAVES

"The Particle Problem in the General Theory of Relativity", by A. Einstein and N. Rosen, *Physical Review*, 1 July 1935, Vol. 48, No. 1, pages 73–77.

"Discovery of a Pulsar in a Binary System", by R.A. Hulse and J.H. Taylor, *The Astrophysical Journal*, 15 January 1975, Vol. 195, No. 2, pages L51–L53.

"Observation of Gravitational Waves from a Binary Black Hole Merger", by B.P. Abbott et al., *Physical Review Letters*, 12 February 2016, Vol. 116, No. 6, pages 061102-1–061102-16.

"Physicists Detect Gravitational Waves", by Andrew Grant, *Science News*, 5 March 2016, Vol. 189, No. 5, pages 6–7.

"Cosmic Shake-Up", by Christopher Crockett, *Science News*, 5 March 2016, Vol. 189, No. 5, pages 22–23.

"Listening for Gravity Waves", by Marcia Bartusiak, *Science News*, 5 March 2016, Vol. 189, No. 5, pages 24–27.

11 THE HEIGHT OF GOOD HEALTH

"Genetic and Environmental Contributions to the Association between Body Height and Educational Attainment: A Study of Adult Finnish Twins", by K. Silventoinen et al., *Behavior Genetics*, November 2000, Vol. 30, No. 6, pages 477–485.

"From the Tallest to (One of) the Fattest: The Enigmatic Fate of the American Population in the 20th Century", by John Komlos and Marieluise Baur, *Economics & Human Biology*, March 2004, Vol. 2, No. 1, pages 57–74.

"An Evaluation of the Relationship between Adult Height and Health-Related Quality of Life in the General UK Population", by T.L. Christensen et al., *Clinical Endocrinology*, September 2007, Vol. 67, No. 3, pages 407–412.

"Underperformance in Affluence: The Remarkable Relative Decline in US Heights in the Second Half of the 20th Century", by John Komlos and Benjamin E. Lauderdale, *Social Science Quarterly*, June 2007, Vol. 88, No. 2, pages 283–305.

"The Recent Decline in the Height of African-American Women", by John Komlos, *Economics & Human Biology*, March 2010, Vol. 8, No. 1, pages 58–66.

"Does Size Matter in Australia?", by Michael Kortt and Andrew Leigh, *Economic Record*, March 2010, Vol. 86, No. 272, pages 71–83.

"Technology Advances; Humans Supersize", by Patricia Cohen, *The New York Times*, 26 April 2011.

"How Have Europeans Grown So Tall?", by Timothy J. Hatton, *Oxford Economic Papers*, April 2014, Vol. 68, No. 2, pages 349–372.

"Defining the Role of Common Variation in the Genomic and Biological Architecture of Adult Human Height", by Andrew R. Wood et al., *Nature Genetics*, November 2014, Vol. 46, No. 11, pages 1173–1186.

"Does Natural Selection Favor Taller Stature Among the Tallest People on Earth?", by Gert Stulp et al., *Proceedings of the Royal Society B*, May 2015, Vol. 282, No. 1806, http://rspb.royalsocietypublishing.org/content/royprsb/282/1806/20150211.full.pdf

12 PYTHON THE CRUSHER

"Boas Know When to Ditch Their Squeeze", *ABC Science*, 18 January 2012, http://www.abc.net.au/science/articles/2012/01/18/3410519.htm

"Snake Modulates Constriction in Response to Prey's Heartbeat", by Scott M. Boback et al., *Biology Letters*, 23 June 2012, Vol. 8, No. 3, pages 473–476.

"Crushing Snakes Kill by Blood Constriction, Not Suffocation", by Kathryn Knight, *The Journal of Experimental Biology*, July 2015, Vol. 218, No. 14, pages 2143–2144.

"Snake Constriction Rapidly Induces Circulatory Arrest in Rats", by Scott M. Boback et al., *The Journal of Experimental Biology*, July 2015, Vol. 218, No. 14, pages 2279–2288.

13 FLY EYES AND SOLAR PANELS

"Natural Photonic Engineers" by Andrew R. Parker, *Materials Today*, 1 September 2002, Vol. 5, No. 9, pages 26–31.

"A Geological History of Reflecting Optics", Andrew Richard Parker, *Journal of the Royal Society Interface*, 22 March 2005, Vol. 2, No. 2, pages 1–17.

"Biomimetics of Photonic Nanostructures", by Andrew R. Parker and Helen E. Townley, *Nature Nanotechnology*, June 2007, Vol. 2, No. 6, pages 347–353.

"Nature's Template", by Andrew Parker, *Chemistry World*, September 2007, Vol. 4, No. 9, pages 54–58.

"Biomimetics: Design by Nature", by Tom Mueller, *National Geographic*, April 2008, http://ngm.nationalgeographic.com/2008/04/biomimetics/tom-mueller-text/1

"Copying Nature's Tricks to Combat Harsh Environments", *The Science Show*, 17 October 2015, http://www.abc.net.au/radionational/programs/scienceshow/copying-naturee28099s-tricks-to-combat-harsh-environments/6860650

14 CEMENT SHOES

"Were 'Concrete Shoes' A Favoured Technique Of Mob Hitmen?", by Cecil Adams, *The Straight Dope*, 14 November 2008, http://www.straightdope.com/columns/read/2824/were-concrete-shoes-a-favored-technique-of-mob-hitmen

"Cement Shoes, Fabled Anchor to Watery Grave, Surface in Brooklyn", by Michael Wilson, *The New York Times*, 4 May 2016, http://www.nytimes.com/2016/05/05/nyre-gion/cement-shoes-fabled-anchor-to-watery-grave-surface-on-body-in-brooklyn.html

"'Cement Shoes' Found on NYC Corpse", *BBC News*, 5 May 2016, http://www.bbc.com/news/world-us-canada-36215804

15 HOW MANY CELLS IN YOUR BODY?

"An Estimation of the Number of Cells in the Human Body", by Eva Bianconi et al., *Annals of Human Biology*, 5 July 2013, Vol. 40, No. 6, pages 463–471.

"How Many Cells Are in Your Body?", by Carl Zimmer, 23 October 2013, *National Geographic Phenomena* blog, http://phenomena.nationalgeographic.com/2013/10/23/how-many-cells-are-in-your-body/

"There Are 37.2 Trillion Cells in Your Body", by Rose Eveleth, 24 October 2013, *Smithsonian.com*, http://www.smithsonianmag.com/smart-news/there-are-372-trillion-cells-in-your-body-4941473/

"37.2 Trillion: Galaxies or Human Cells?", by Nicholas Bakalar, *The New York Times*, 19 June 2015, http://www.nytimes.com/2015/06/23/science/37-2-trillion-galaxies-or-human-cells.html

"How Many Cells Are There In The Human Body?", by Robin Corey, *Quora*, 22 July 2015, https://www.quora.com/How-many-cells-are-there-in-the-human-body

"How Many Cells Are There in the Human Body – and How Many Microbes?", by Michael Greshko, *National Geographic*, 13 January 2016, http://news.nationalgeographic.com/2016/01/160111-microbiome-estimate-count-ratio-human-health-science/

"Revised Estimates for the Number of Human and Bacteria Cells in the Body", by Ron Sender et al., 6 January 2016, *bioRxiv*, http://biorxiv.org/content/early/2016/01/06/036103

"Body's Bacteria Don't Outnumber Human Cells So Much After All", by Tina Hesman Saey, *ScienceNews*, 6 February 2016, Vol. 189, No. 3, page 6.

16 WATER BURNS PLANTS?

"Watering Plants Midday Triggers Sunburn, Research Shows", *European Union Community Research and Development Information Service*, 26 January 2010, http://cordis.europa.eu/news/rcn/31699_en.html

"Can Water Droplets on Leaves Cause Leaf Scorch?", by Hamlyn G. Jones, *New Phytologist*, March 2010, Vol. 185, No. 4, pages 865–867.

"Optics of Sunlit Water Drops on Leaves: Conditions Under Which Sunburn Is Possible", by Adam Egri et al., *New Phytologist*, March 2010, Vol. 185, No. 4, pages 979–987.

"Sunburnt Plants 'Myth' Is Debunked", by Richard Gray, *Telegraph* (UK), 13 June 2010, http://www.telegraph.co.uk/news/science/science-news/7823032/Sunburnt-plants-myth-is-debunked.html

17 DIRTY DATA

"Information Wants to Be Free . . . and Expensive", by Jennifer Lai, 20 July 2009, *Fortune*, http://fortune.com/2009/07/20/information-wants-to-be-free-and-expensive/

"How Dirty is Your Data: A Look at the Energy Choices That Power Cloud Computing", by Gary Cook and Jodie Van Horn, *Greenpeace International*, April 2011, http://www.greenpeace.org/international/Global/international/publications/climate/2011/Cool%20IT/dirty-data-report-greenpeace.pdf/

"The Rise and Fall of Bitcoin", by Benjamin Wallace, *Wired*, 23 November 2011, http://www.wired.com/2011/11/mf_bitcoin/

"Data Barns in a Farm Town, Gobbling Power and Flexing Muscle", by James Glanz, *The New York Times*, 23 September 2012, http://www.nytimes.com/2012/09/24/technology/data-centers-in-rural-washington-state-gobble-power.html?_r=0

"Virtual Bitcoin Mining is a Real-World Environmental Disaster", by Mark Gimein, *Bloomberg*, 13 April 2013, http://www.bloomberg.com/news/articles/2013-04-12/virtual-bitcoin-mining-is-a-real-world-environmental-disaster

"Netflix Gobbles a Third of Peak Internet Traffic in North America", by Don Reisinger, *CNET*, 7 November 2012, http://www.cnet.com/news/netflix-gobbles-a-third-of-peak-internet-traffic-in-north-america/

"Global Bitcoin Computing Power Now 256 Times Faster Than Top 500 Supercomputers, Combined!", by Reuven Cohen, *Forbes*, 28 November 2013, http://www.forbes.com/sites/reuvencohen/2013/11/28/global-bitcoin-computing-power-now-256-times-faster-than-top-500-supercomputers-combined/#6a719dae28b7

"Clicking Clean: A Guide to Building the Green Internet", by Gary Cook and David Pomerantz, *Greenpeace Inc.*, May 2015, http://www.greenpeace.org/usa/wp-content/uploads/legacy/Global/usa/planet3/PDFs/2015ClickingClean.pdf

"Guzzling Data: Australian Internet Downloads Explode", by Hannah Francis, *The Sydney Morning Herald*, 1 April 2015, http://www.smh.com.au/digital-life/digital-life-news/guzzling-data-australian-internet-downloads-explode-20150401-1mcqgq.html

"The Environmental Toll of a Netflix Binge", by Ingrid Burrington, *The Atlantic*, 16 December 2015, http://www.theatlantic.com/technology/archive/2015/12/there-are-no-clean-clouds/420744/

"The Dirty Parts of the Computing World", by Nathan Ensmenger, *Bulletin of the Atomic Scientists*, 11 April 2016, http://thebulletin.org/dirty-parts-computing-world9312

The World's Most Powerful Computer Network is Being Wasted on Bitcoin", by Eric Limer, *Gizmodo*, 31 May 2016, http://gizmodo.com/the-worlds-most-powerful-computer-network-is-being-was-504503726

18 IMMORTAL JELLYFISH

"Bi-Directional Conversion in *Turritopsis nutricula* (Hydrozoa)", by Giorgio Bavestrollo et al., *Scientia Marina*, 1992, Vol. 56, No. 2–3, pages 137–140.

"Reversing the Life Cycle: Medusae Transforming into Polyps and Cell Transdifferentiation in *Turritopsis nutricula* (Cnidaria, Hydroza)", by Stefano Piraino et al., *Biological Bulletin*, June 1996, Vol. 190, No. 3, pages 302–312.

"Evidence of Reverse Development in *Leptomedusae* (Cnidaria, Hydrozoa): The Case of *Laodicea undulata* (Forbes and Goodsir 1851)", by D. De Vito et al., *Marine Biology*, May 2006, Vol. 149, No. 2, pages 339–346.

"Species in the Genus *Turritopsis* (Cnidaria, Hydrozoa): A Molecular Evaluation", by M.P. Miglietta et al., *Journal of Zoological Systematics and Evolutionary Research*, February 2007, Vol. 45, No. 1, pages 11–19.

"*Turritopsis nutricula*", by Hongbao Ma and Yan Yang, *Nature and Science*, 2010, Vol. 8, No. 2, pages 15–20.

"Forever and Ever", Nathaniel Rich, *The New York Times Magazine*, 28 November 2012.

"Life Cycle Reversal in *Aurelia* sp.1 (Cnidaria, Scyphozoa)", by Jinru He et al., *PLoS One*, 21 December 2015, http://journals.plos.org/plosone/article?id=10.1371/journal.pone.0145314

"The Immortal Jellyfish: Researchers Find Creature Can Age Backwards, Form Hordes of Clones and Regenerate Lost Body Parts", by Ellie Zolfagharifard, *Daily Mail*, 3 March 2016, http://www.dailymail.co.uk/sciencetech/article-3473857/The-immortal-jellyfish-Researchers-creature-age-backward-form-hordes-clones-regenerate-lost-body-parts.html

19 HOT TEA COOLS YOU DOWN

"Does Drinking Tea on a Hot Day Cool You Down?", by Robert Innes, *Sennir*, 26 August 2007, http://www.sennir.co.uk/Journal/Does_Tea_Cool_You_Down

"A Hot Drink on a Hot Day Can Cool You Down", by Joseph Stromberg, *Smithsonian*, 10 July 2012, http://www.smithsonianmag.com/science-nature/a-hot-drink-on-a-hot-day-can-cool-you-down-1338875/?no-ist

"Cool Down with a Hot Drink? It's Not as Crazy as You Think", by Joe Palca, *NPR Morning Edition*, 11 July 2012, http://www.npr.org/sections/the-salt/2012/07/11/156378713/cool-down-with-a-hot-drink-its-not-as-crazy-as-you-think

"Does Drinking Hot Drinks on a Scorching Summer's Day Really Cool You Down?", Jessie Wingard, *DW*, 25 July 2013, http://www.dw.com/en/does-drinking-hot-drinks-on-a-scorching-summers-day-really-cool-you-down/a-16974502

"A Little Warmth Goes a Long Way – The Science of Hot Drinks", by Amy Fleming, *The Guardian*, 28 October 2014, http://www.theguardian.com/lifeandstyle/wordofmouth/2014/oct/28/hot-drinks-science-tasting-notes

"Ice Cream Won't Keep You Cool, But Cup of Tea Will", Rachel Clun, *The Sydney Morning Herald*, 14 January 2015, http://www.smh.com.au/lifestyle/diet-and-fitness/ice-cream-wont-keep-you-cool-but-cup-of-tea-will-20150112–12n0kl.html

"Fact Buster: Can a Hot Drink Help Keep You Cool in Hot Weather?", by Nicola Garrett, *ABC Health and Wellbeing*, 2 February 2015, http://www.abc.net.au/health/talkinghealth/factbuster/stories/2015/02/05/4173208.htm

20 TIME TRAVEL

"Around-the-World Atomic Clocks: Predicted Relativistic Time Gains", by J.C. Hafele and Richard E. Keating, *Science*, 14 July 1972, Vol. 177, No. 4044, pages 166–168.

"Five Ways to Travel Through Time", by Cathal O'Connell, *Cosmos*, 5 April 2016, https://cosmosmagazine.com/physics/five-ways-travel-through-time

21 BITCOIN: LEGEND OF A LEDGER

"Bitcoin: A Peer-to-Peer Electronic Cash System", by Satoshi Nakamoto, *Bitcoin.org*, 1 November 2008, https://bitcoin.org/bitcoin.pdf

"Pizza for Bitcoins?", *Bitcoin Forum*, 18 May 2010, https://bitcointalk.org/index.php?topic=137.0

"The Rise and Fall of Bitcoin", by Benjamin Wallace, *Wired*, 23 November 2011, http://www.wired.com/2011/11/mf_bitcoin/

"Geeks Love the Bitcoin Phenomenon Like They Loved the Internet in 1995", by Ken Tindell, 6 April 2013, *Business Insider*, http://www.businessinsider.com.au/how-bitcoins-are-mined-and-used-2013-4

"Virtual Bitcoin Mining Is a Real-World Environmental Disaster", by Mark Gimein, *Bloomberg*, 13 April 2013, http://www.bloomberg.com/news/articles/2013-04-12/virtual-bitcoin-mining-is-a-real-world-environmental-disaster

"Optimizing Bitcoin Generation and the Feasibility of Profitability", by Michael Sprague, UC Santa Barbara, June 2013, http://www.cs.ucsb.edu/~msprague/bitcoin_profitability_final.pdf

"Global Bitcoin Computing Power Now 256 Times Faster than Top 500 Supercomputers, Combined!", by Reuven Cohen, *Forbes*, 28 November 2013, http://www.forbes.com/sites/reuvencohen/2013/11/28/global-bitcoin-computing-power-now-256-times-faster-than-top-500-supercomputers-combined/#6a719dae28b7

"Bitcoin Now Accepted by 100,000 Merchants Worldwide", by Anthony Cuthbertson, *International Business Times*, 4 February 2015, http://www.ibtimes.co.uk/bitcoin-now-accepted-by-100000-merchants-worldwide-1486613

"A Living Currency", 22 May 2015, http://bitcoinwhoswho.com/index/jercosinterview

"Bitcoin Basics" by Nathaniel Popper, *The New York Times* Business Day, 4 November 2015, http://www.nytimes.com/2015/11/05/business/bitcoin-basics.html

"The Resolution of the Bitcoin Experiment" by Mike Hearn, *Medium*, 15 January 2016, https://medium.com/@octskyward/the-resolution-of-the-bitcoin-experiment-dabb30201f7#.5646tiuzm

"When Bitcoin Grows Up", by John Lancaster, *London Review of Books*, 21 April 2016, Vol. 38, No. 8, pages 3–12, http://www.lrb.co.uk/v38/n08/john-lanchester/when-bitcoin-grows-up

"Craig Wright's New Evidence that He Is Satoshi Nakamoto Is Worthless", by Jordan Pearson and Lorenzo Franceschi-Bicchierai, 2 May 2016, *Vice Motherboard*, http://motherboard.vice.com/read/craig-wright-satoshi-nakamoto-evidence-signature-is-worthless

"The World's Most Powerful Computer Network Is Being Wasted on Bitcoin", by Eric Limer, *Gizmodo*, 31 May 2016, http://gizmodo.com/the-worlds-most-powerful-computer-network-is-being-was-504503726

"The Bitcoin Volatility Index", accessed 12 June 2016, https://btcvol.info

22 SPLEEN AND RED BLOOD CELLS

"Scavenger Receptors on Liver Kupffer Cells Mediate the In Vivo Uptake of Oxidatively Damaged Red Blood Cells in Mice", by Valeska Terpstra and Theo J.C. van Berkel, *Blood*, 15 March 2000, Vol. 95, No. 6, pages 2157–2163.

"Finally, the Spleen Gets Some Respect", by Natalie Angier, *The New York Times*, 3 August 2009.

"Dispensable But Not Irrelevant", by Ting Jia and Eric G. Pamer, *Science*, 31 July 2009, Vol. 325, No. 5940, pages 549–550.

"Identification of Splenic Reservoir Monocytes and Their Deployment to Inflammatory Sites", by Filip K. Swirski et al., *Science*, 31 July 2009, Vol. 325, No. 5940, pages 612–616.

"Physiology and Pathophysiology of Eryptosis", by Florian Lang et al., *Transfusion Medicine and Hemotherapy*, October 2012, Vol. 39, No. 5, pages 308–314.

23 MOVIE AUDIENCES EMIT CHEMICALS

"Cinema Audiences Reproducibly Vary the Chemical Composition of Air During Films, by Broadcasting Scene Specific Emissions on Breath", by Jonathan Williams et al., *Scientific Reports*, 10 May 2016, Vol. 6, No. 25464, http://www.nature.com/articles/srep25464

"Audience Reactions Alter Cinema Atmosphere", by Anthony King, *Chemistryworld*, 20 May 2016, http://www.rsc.org/chemistryworld/2016/05/cinema-audience-atmosphere-isoprene-volatiles

24 MOZZIES LOVE (SOME) HUMANS

"The Claim: Mosquitoes Attack Some People More Than Others", by Anahad O'Connor, *The New York Times*, 21 June 2005, http://www.nytimes.com/2005/06/21/health/the-claim-mosquitoes-attack-some-people-more-than-others.html

"The Claim: Eating Garlic Helps Repel Mosquitoes", by Anahad O'Connor, *The New York Times*, 24 July 2007.

"The Claim: Mosquitoes Are Attracted to Women More Than to Men", by Anahad O'Connor, *The New York Times*, 15 June 2010.

"Are Some People More Attractive to Mosquitoes?", by Jenny Pogson, *ABC Health & Wellbeing*, 30 November 2011, http://www.abc.net.au/health/talkinghealth/factbuster/stories/2011/11/30/3379114.htm

"Monday's Medical Myth: Mosquitos Prefer Sweet Blood", by Cameron Webb, *The Conversation*, 10 December 2012, https://theconversation.com/mondays-medical-myth-mosquitos-prefer-sweet-blood-10833

"Mosquitoes, Beer & Australia Day", by Cameron Webb, *Mosquito Research and Management*, 25 January 2013, https://cameronwebb.wordpress.com/2013/01/25/mosquitoes-beer-australia-day/

"Targeting a Dual Detector of Skin and CO_2 to Modify Mosquito Host Seeking", by Genevieve M. Tauxe et al., *Cell*, 5 December 2013, Vol. 155, No. 6, pages 1365–1379.

"Health Check: Why Mosquitoes Seem to Bite Some People More", by Cameron Webb, *The Conversation*, 26 January 2015, https://theconversation.com/health-check-why-mosquitoes-seem-to-bite-some-people-more-36425

"Heritability of Attractiveness to Mosquitoes", by G. Mandela Fernández-Grandon et al., *PLoS One*, 22 April 2015, http://journals.plos.org/plosone/article?id=10.1371/journal.pone.0122716

"Mozzie Magnetism is 'All in the Genes'", by Bianca Nogrady, *ABC Science*, 23 April 2015, http://www.abc.net.au/science/articles/2015/04/23/4221543.htm

25 DOUBLE-YOLK EGGS

"What Are The Chances of Six Double-Yolkers?", by Wesley Stephenson, *BBC Radio 4 More or Less*, 10 December 2011, http://www.bbc.com/news/magazine-16118149

"Double-Yolk Guarantee: M&S Cracks Formula for Boxes of Super-Eggs (If You'll Shell Out £2.75)", by Sean Poulter, *Daily Mail*, 7 February 2015, http://www.dailymail.co.uk/news/article-2943313/M-S-cracks-formula-boxes-super-eggs-guarantee-double-yolks.html

27 CREDIT CARD THEFT

"Identity Theft Vs. Credit Card Fraud", *NPR Morning Edition*, 18 July 2005, http://www.npr.org/templates/story/story.php?storyId=4758474

"Would You Have Spotted the Fraud?", by Brian Krebs, *Krebs on Security*, 15 January 2010, http://krebsonsecurity.com/2010/01/would-you-have-spotted-the-fraud/

"5 Ways Thieves Steal Credit Card Data", by Janna Herron, *Bankrate.com*, 15 August 2011, http://www.bankrate.com/finance/credit-cards/5-ways-thieves-steal-credit-card-data-1.aspx

"Stolen Credit Cards and the Black Market: How the Digital Underground Works", *Tripwire: The State of Security*, 21 December 2013, http://www.tripwire.com/state-of-security/vulnerability-management/how-stolen-target-credit-cards-are-used-on-the-black-market/

"The Underground Economy of Data Breaches", by Wade Williamson, *Forbes*, 18 June 2014, http://www.forbes.com/sites/frontline/2014/06/18/the-underground-economy-of-data-breaches/#45e94b936c72

"What Is the Dark Net, and How Will It Shape the Future of the Digital Age?", by Steven Viney, *ABC News*, 27 January 2016, http://www.abc.net.au/news/2016-01-27/explainer-what-is-the-dark-net/7038878

"Hackers' $81 Million Sneak Attack on World Banking", by Michael Corkery, *The New York Times*, 30 April 2016, http://www.nytimes.com/2016/05/01/business/dealbook/hackers-81-million-sneak-attack-on-world-banking.html

28 SLEEP BADLY IN AN UNFAMILIAR BED

"Left Brain Stands Guard While Sleeping Away from Home", by Laura Sanders, *Science News*, 21 April 2016, https://www.sciencenews.org/article/left-brain-stands-guard-while-sleeping-away-home

"Neurological Night Watch: Why a Familiar Bed Provides a Good Night's Sleep", *The Economist*, 23 April 2016, http://www.economist.com/news/science-and-technology/21697213-why-familiar-bed-provides-good-nights-sleep-neurological-night-watch

"Night Watch in One Brain Hemisphere During Sleep Associated with the First-Night Effect in Humans", by Masako Tamaki et al., *Current Biology*, 9 May 2016, Vol. 26, No. 9, pages 1190–1194.

29 SMOOTHIE SCAM

"Smoothies: One Portion or Two?", by C.H.S. Ruxton, *British Nutrition Foundation Nutrition Bulletin*, June 2008, Vol. 33, No. 2, pages 129–132.

"7 Health Fads from 2014", by Cassie White, *ABC Health and Wellbeing*, 18 December 2014, http://www.abc.net.au/health/features/stories/2014/12/18/4150272.htm

"Ask Well: The Downside of Smoothies", by Roni Caryn Rabin, *The New York Times Well* blog, 13 May 2016, http://well.blogs.nytimes.com/2016/05/13/ask-well-the-downside-of-smoothies/

30 SUNSCREEN ATTACKS CORAL REEF

"Swimmers' Sunscreen Killing Off Coral", by Ker Than, *National Geographic News*, 29 January 2008, http://news.nationalgeographic.com/news/2008/01/080129-sunscreen-coral.html

"The New Rules for Sunscreen", by Roni Caryn Rabin, *The New York Times Well* blog, 27 May 2013, http://well.blogs.nytimes.com/2013/05/27/the-new-rules-for-sunscreen/

"Sunscreen Contributing to Decline of Coral Reefs, Study Shows", by Reuters, *The Guardian*, 21 October 2015, https://www.theguardian.com/environment/2015/oct/21/sunscreen-contributing-to-decline-of-coral-reefs-study-shows

"Sunscreen Chemical Oxybenzene Blamed for Harm to Coral Reefs", by AFP, *ABC News*, 22 October 2015, http://www.abc.net.au/news/2015-10-22/sunscreen-chemical-blamed-for-harm-to-coral-reefs/6876036

"Toxicopathological Effects of the Sunscreen UV Filter, Oxybenzone (Benzophenone-3), on Coral Planula and Cultured Primary Cells and Its Environmental Contamination in Hawaii and the US Virgin Islands", by C.A. Downs et al., *Archives of Environmental Contamination and Toxicology*, February 2016, Vol. 70, No. 2, pages 265–288.

31 COFFEE'S A DIURETIC?

"Tolerance and Cross-Tolerance in the Human Subject to the Diuretic Effect of Caffeine, Theobromine and Theophylline", by Nathan B. Eddy and Ardrey W. Downs, *Journal of Pharmacology and Experimental Therapeutics*, June 1928, Vol. 33, No. 2, pages 167–174.

"Coffee Consumption and Total Body Homoeostasis as Measured by Fluid Balance and Biological Impedance Analysis", by M. Neuhäuser-Berthold et al., *Annals of Nutrition and Metabolism*, January 1997, pages 29–36.

"The Effect of Caffeinated, Non-Caffeinated, Caloric and Non-Caloric Beverages on Hydration", by A.C. Grandjean et al., *Journal of the American College of Nutrition*, October 2000, Vol. 19, No. 5, pages 591–600.

"Caffeine, Body Fluid-Electrolyte Balance and Exercise Performance" by Lawrence E. Armstrong, *International Journal of Sport Nutrition and Exercise Metabolism*, June 2002, Vol. 12, No. 2, pages 189–206.

"Black Tea Is Not Significantly Different from Water in the Maintenance of Normal Hydration in Human Subjects: Results from a Randomised Controlled Trial", by Carrie H. Ruxton and Valerie A. Hart, *British Journal of Nutrition*, August 2011, Vol. 106, No. 4, pages 588–595.

"No Evidence of Dehydration with Moderate Daily Coffee Intake: A Counterbalanced Cross-Over Study in a Free-Living Population", by Sophie C. Killer et al., *PLoS One*, January 2014, e84154.

"Do Coffee and Tea Really Dehydrate Us?", by Claudia Hammond, *BBC Future*, 2 April 2014, http://www.bbc.com/future/story/20140402-are-coffee-and-tea-dehydrating

"Caffeine and Diuresis during Rest and Exercise: A Meta-Analysis", by Yang Zhang et al., *Journal of Science in Medicine and Sport*, September 2015, Vol. 18, No. 5, pages 569–574.

32 DRAGONFLY TELESCOPE

"Ultra Low Surface Brightness Imaging with the Dragonfly Telephoto Array", by Roberto G. Abraham and Pieter G. van Dokkum, arXiv.org, 21 Jan 2014, arXiv:1401.5473v1 [astro-ph.IM].

"Forty-Seven Milky Way-Sized, Extremely Diffuse Galaxies in the Coma Cluster", by Pieter G. van Dokkum et al., The Astrophysical Journal Letters, 10 January 2015, Vol. 798, No. 2, pages 1–8.

"Ghostly Galaxies Appear in the Coma Cluster", by Ken Croswell, Scientific American, 1 April 2015, http://www.scientificamerican.com/article/ghostly-galaxies-appear-in-the-coma-cluster/

"How to Discover a Galaxy with a Telephoto Lens", by Patchen Barss, Nautilus, 28 January 2016, http://nautil.us/issue/32/space/how-to-discover-a-galaxy-with-a-telephoto-lens

"The Mystery of Phantom Galaxies May Soon Be Solved", by Ken Croswell, Scientific American, 22 April 2016, http://www.scientificamerican.com/article/the-mystery-of-phantom-galaxies-may-soon-be-solved/

33 PHUNDAMENTAL PHYSICS PROBLEMS

"The Hierarchy Problem", by Matt Strassler, *Of Particular Significance*, 14 August 2011, https://profmattstrassler.com/articles-and-posts/particle-physics-basics/the-hierarchy-problem/

"Have We Reached the End of Physics?", by Harry Cliff, *TEDGlobal>Geneva*, December 2015, http://www.ted.com/talks/harry_cliff_have_we_reached_the_end_of_physics/transcript?language=en

"The Two Most Dangerous Numbers in the Universe Could Signal the End of Physics", by Jessica Orwig, *Business Insider*, 15 January 2016, http://www.sciencealert.com/the-2-most-dangerous-numbers-in-the-universe-could-signal-the-end-of-physics

General Relativity: An Introduction for Physicists, by M.P. Hobson, G.P. Efstathios and A.N. Lasenby, Cambridge University Press, Cambridge, 2006, page 187.

34 PLANET – A GIANT BABY

"Hubble Directly Observes Planet Orbiting Fomalhaut", *Hubble Space Telescope*, 13 November 2008, http://www.spacetelescope.org/news/heic0821/

"Astronomers Observe Newborn Planets Evolving from Gas and Dust Particles", by Stuart Gary, *ABC News*, 19 November 2015, http://www.abc.net.au/news/2015–11–19/planetary-formation-seen-for-the-first-time-by-astronomers/6951494

"Accreting Protoplanets in the LkCa 15 Transition Disk", by S. Sallum et al., *Nature*, 19 November 2015, Vol. 527, No. 7578, pages 342–344.

"How Two Tiny Dots Defy the History of Life and the Solar System", by Colin Stuart, *New Scientist*, 23 April 2016, pages 30–33.

35 POISONED PANTS

"Paraquat", *Pesticides News*, June 1996, pages 20–21, http://www.pan-uk.org/pestnews/Actives/paraquat.htm

"Paraquat and Suicide: Fact Sheet", PAN Germany, 2003, http://www.pan-germany.org/download/fact_paraquat2.pdf

"Material Safety Data Sheet: eChem Paraquat 250 Herbicide", eChem Australia, November 2013, http://www.echem.com.au/pdf/Herbicides/Paraquat/Paraquat-MSDS.pdf

"Underwear 'Soaked' in Poison 'Rots Man's Genitals'", by Toby Meyjes, *Metro.co.uk*, 10 May 2016, http://metro.co.uk/2016/05/10/underwear-soaked-in-poison-rots-mans-genitals-5873674/

"Knickers in a Twist: The Case of the Poisoned Pants", by Kathryn Harkup, *The Guardian*, 27 May 2016, https://www.theguardian.com/science/blog/2016/may/27/knickers-in-a-twist-the-case-of-the-poisoned-pants-paraquat

36 I, VOMIT

"Were There Really Vomitoriums in Ancient Rome?", by Cecil Adams, *The Straight Dope*, 1 November 2002, http://www.straightdope.com/columns/read/2421/were-there-really-vomitoriums-in-ancient-rome

"Vomitoriums: Fact or Fiction?", by Stephanie Butler, *Hungry History*, 20 November 2012, http://www.history.com/news/hungry-history/vomitoriums-fact-or-fiction

37 VOMITING MACHINE

"Hand Washing Vs. Hand Sanitiser in the Fight against Norovirus", by Barry Michaels, *Debgroup.com*, 19 February 2014, http://info.debgroup.com/blog/bid/336078/Hand-Washing-vs-Hand-Sanitizer-in-the-Fight-Against-Norovirus

"'Vomiting Device' Sounds Gross But It Helps Study Infections", by Sarah Schwartz, *ScienceNews*, 19 September 2015, Vol. 188, No. 6, page 5.

"Aerosolization of a Human Norovirus Surrogate, Bacteriophage MS2, During Simulated Vomiting", by Grace Tung-Thompson et al., *PLoS One*, 19 August 2015, http://journals.plos.org/plosone/article?id=10.1371/journal.pone.0134277

38 PLACES OF PI

"The Mountains of Pi", by Richard Preston, *The New Yorker*, 2 March 1992.

"A History of Pi", by J.J. O'Connor and E.F. Robertson, School of Mathematics and Statistics, University of St Andrews, Scotland, August 2001, http://www-groups.dcs.st-andrews.ac.uk/~history/HistTopics/Pi_through_the_ages.html

"Johann Martin Zacharias Dase", by J.J. O'Connor and E.F. Robertson, School of Mathematics and Statistics, University of St Andrews, Scotland, July 2009, http://www-groups.dcs.st-and.ac.uk/history/Biographies/Dase.html

"Take It to the Limit", by Steven Strogatz, *The New York Times Opinionator* blog, 4 April 2010, http://opinionator.blogs.nytimes.com/2010/04/04/take-it-to-the-limit/

"New Math: The Time Indiana Tried to Change Pi to 3.2", by Ethan Trex, *Mental Floss*, 14 March 2016, http://mentalfloss.com/article/30214/new-math-time-indiana-tried-change-pi-32

"Life of 3.14159", by Bernie Hobbs, *ABC Science*, 13 March 2013, http://www.abc.net.au/science/articles/2013/03/13/3714326.htm

"Counting the Yarrow: An 18th Century Method for Calculating Pi", by Jennifer Ouellette, *Scientific American Cocktail Party Physics* blog, 14 March 2013, http://blogs.scientificamerican.com/cocktail-party-physics/casting-the-yarrow-an-18th-century-method-for-calculating-pi/

"A Ballistic Monte Carlo Approximation of π", by Vincent Dumoulin and Félix Thouin, 8 April 2014, arXiv:1404.1499v2 [physics.pop-ph], http://arxiv.org/pdf/1404.1499v2.pdf

"How Mathematicians Used a Pump-Action Shotgun to Estimate Pi", *The Physics arXiv Blog*, 13 April 2014, https://medium.com/the-physics-arxiv-blog/how-mathematicians-used-a-pump-action-shotgun-to-estimate-pi-c1eb776193ef#.4kktow49j

"How to Calculate Pi Using a Pump-Action Shotgun", by Michael Byrne, *Vice Motherboard* blog, 4 January 2015, http://motherboard.vice.com/read/how-to-calculate-pi-using-a-pump-action-shotgun

"How Many Decimals of Pi Do We Really Need", *NASA/JPL Edu*, 16 March 2016, http://www.jpl.nasa.gov/edu/news/2016/3/16/how-many-decimals-of-pi-do-we-really-need/

39 ELECTRIC MOTORS IN BACTERIA

"An Introduction to the Physics of the Bacterial Flagellar Motor: a Nanoscale Rotary Electric Motor", by Matthew A.B. Baker and Richard M. Berry, Contemporary Physics, November 2009, Vol. 50, No. 6, pages 617–632.

"Temperature Dependences of Torque Generation and Membrane Voltage in the Bacterial Flagellar Motor", by Yuichi Inoue et al., Biophysical Journal, 17 December 2013, Vol. 105, No. 12, pages 2801–2810.

"A Delicate Nanoscale Motor Made by Nature – The Bacterial Flagellar Motor", by Ruidong Xue et al., Advanced Science, September 2015, Vol. 2, No. 9, 1500129.

"Domain-Swap Polymerisation Drives the Self-Assembly of the Bacterial Flagellar Motor", by Matthew A.B. Baker et al., Nature Structural and Molecular Biology, March 2016, Vol. 23, No. 3, pages 197–203.

"High-Power Biological Wheels and Motors Imaged for the First Time", by Rowan Hooper, New Scientist, 14 March 2016, https://www.newscientist.com/article/2080642-high-power-biological-wheels-and-motors-imaged-for-first-time/

"Bacterial Flagellar Motor Switch in Response to CheY-P Regulation and Motor Structural Alterations", by Qi Ma et al., Biophysical Journal, 29 March 2016, Vol. 110, No. 6, pages 1411–1420.

"Diverse High-Torque Bacterial Flagellar Motors Assemble Wider Stator Rings Using a Conserved Protein Scaffold", by Morgan Beeby et al., PNAS, 29 March 2016, pages E1917–E1926.

40 COFFEE IS NOW GOOD

"Prevention: Coffee Lowers Risk of Prostate Cancer, Harvard Study Says", by Roni Caryn Rabin, *The New York Times*, 20 May 2011, http://www.nytimes.com/2011/05/24/health/research/24prevention.html

"Coffee's Cancer Reducing Effect Explained", *ABC Science*, 17 August 2011, http://www.abc.net.au/science/articles/2011/08/17/3295499.htm

"Habitual Coffee Consumption and Risk of Heart Failure: A Dose-Response Meta-Analysis", by Elizabeth Mostofsky et al., *Circulation: Heart Failure*, July 2012, Vol. 5, No. 4, pages 401–405.

"More Consensus on Coffee's Effect on Health than You Might Think", by Aaron E. Carroll, 11 May 2015, *The New York Times*, http://www.nytimes.com/2015/05/12/upshot/more-consensus-on-coffees-benefits-than-you-might-think.html

"Health Check: Four Reasons to Have Another Cup of Coffee", by Clare Collins, *The Conversation*, 25 May 2015, https://theconversation.com/health-check-four-reasons-to-have-another-cup-of-coffee-40390

41 WEIGHT LOSS VIA EXERCISE

"Training in the Fasted State Improves Glucose Tolerance During Fat-Rich Diet", by Karen Van Proeyen et al., *Journal of Physiology*, November 2010, Vol. 588, No. 21, pages 4289–4302.

"Ask Well: The Best Time to Exercise to Lose Weight", by Gretchen Reynolds, *The New York Times Well* blog, 23 January 2015, http://well.blogs.nytimes.com/2015/01/23/ask-well-the-best-time-of-day-to-exercise/

THANK YOU

IF TIME TRAVEL WERE POSSIBLE, MAYBE I COULD USE IT TO TAKE OFF ON FRIDAY FOR A WEEKEND OF SERIOUS PARTYING AND ROCK AND ROLL – AND TURN UP ON MONDAY WITH A BEAUTIFULLY CRAFTED AND ELEGANTLY WRITTEN (AND FACTUALLY CORRECT) BOOK. BUT IT AIN'T SO.

I thank my lovely family who (like the Tardis) manage more than you can see on the outside – Lola, Alice, Little Karl (who is now taller than me), Mary, and Brendan. Only thanks to them could this book be written.

I am not an expert in any field, so let me thank those who are, and who helped me. (Standard Disclaimer – all mistakes are my fault, all goodness is due to them.) So Big It Up for Goncalo Borges and Helen Sim (Bitcoin), Clare Collins (Coffee), Geraint Lewis (Gravitational Waves), Cameron Webb (Mosquitoes and other Insects), and Peter Tuthill (Giant Baby Planet).

But words have to be corralled and captured inside the pages of a paper (and electronic) book. So on the Publishing side, let me thank Claire Craig (publisher), Jo Butler (agent), Danielle Walker and Sarah Fletcher, and Alex Christie (publicity).

The layout/design and illustrations are due to the Good Folk at Xou Creative – Jon MacDonald and Roy Chen.

I deeply thank Isabelle Benton (my Producer at the University of Sydney) and both Dan Driscoll and Tiger Webb (my Producers at the Australian Broadcasting Corporation) – and Mary yet again. They all reshaped the stories – and created most of the punchlines.

Finally, let me thank You, the Mind Traveller (if we can't have Time Travel today, at least Thinking is the Best Way to Travel). Without you, I would have nobody to write to, and for.

Also by
DR KARL KRUSZELNICKI

- *Curiouser & Curiouser: Burping Cows, Bending Spoons, Beer Goggles & other scintillating scientific stories...*
- *Brain Food*
- *50 Shades of Grey Matter*
- *Game of Knowns: Science is Coming*

Also by
DR KARL KRUSZELNICKI

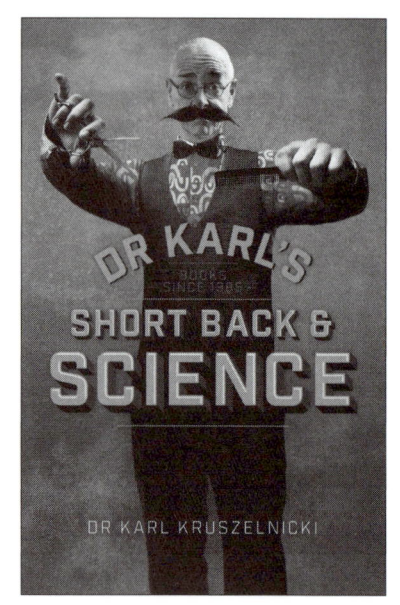